Reteaching Workbook

TAKE ANOTHER LOOK

TEACHER'S EDITION
Grade 7

Harcourt Brace & Company
Orlando • Atlanta • Austin • Boston • San Francisco • Chicago • Dallas • New York • Toronto • London
http://www.hbschool.com

Copyright © by Harcourt Brace & Company

All rights reserved. No part of this publication may be reproduced or transmitted in any form or by any means, electronic or mechanical, including photocopy, recording, or any information storage and retrieval system, without permission in writing from the publisher.

Requests for permission to make copies of any part of the work should be mailed to: Permissions Department, Harcourt Brace & Company, 6277 Sea Harbor Drive, Orlando, Florida 32887-6777.

HARCOURT BRACE and Quill Design is a registered trademark of Harcourt Brace & Company. MATH ADVANTAGE is a trademark of Harcourt Brace & Company.

Printed in the United States of America

ISBN 0-15-311077-5

1 2 3 4 5 6 7 8 9 10 073 2000 99 98

CONTENTS

CHAPTER 1 — Making Number Connections
- 1.1 Sets of Numbers R1
- 1.2 Understanding Rational Numbers R2
- 1.3 Parts as Percents R3
- 1.4 Connecting Fractions, Decimals, and Percents R4
- 1.5 Making Circle Graphs R5

CHAPTER 2 — Expressing Numbers
- 2.1 Using Exponents R6
- 2.2 Exploring Decimal and Binary Numbers R7
- 2.3 Modeling Squares and Square Roots R8
- 2.4 Problem-Solving Strategy: Using Guess and Check to Find Square Roots R9
- 2.5 Repeated Calculations R10

CHAPTER 3 — Using Whole Numbers and Decimals
- 3.1 Estimating Sums and Differences R11
- 3.2 Multiplying Whole Numbers and Decimals R12
- 3.3 Dividing Whole Numbers and Decimals R13
- 3.4 Order of Operations R14

CHAPTER 4 — Operations with Fractions
- 4.1 Adding and Subtracting Fractions R15
- 4.2 Adding and Subtracting Mixed Numbers R16
- 4.3 Estimating Sums and Differences R17
- 4.4 Multiplying and Dividing Fractions and Mixed Numbers R18
- 4.5 Problem-Solving Strategy: Solving a Simpler Problem R19

CHAPTER 5 — Operations with Integers and Rational Numbers
- 5.1 Adding Integers R20
- 5.2 Subtracting Integers R21
- 5.3 Multiplying and Dividing Integers R22
- 5.4 Adding and Subtracting Rational Numbers R23
- 5.5 Multiplying and Dividing Rational Numbers R24

CHAPTER 6 — Writing and Simplifying Expressions
- 6.1 Numerical and Algebraic Expressions R25
- 6.2 Evaluating Expressions R26
- 6.3 Combining Like Terms R27
- 6.4 Sequences and Expressions R28

CHAPTER 7 — Solving One-Step Equations
- 7.1 Connecting Equations and Words R29
- 7.2 Solving Addition and Subtraction Equations R30
- 7.3 Multiplication and Division Equations R31
- 7.4 Problem-Solving Strategy: Working Backward to Solve Problems R32
- 7.5 Proportions R33

CHAPTER 8 — Solving Two-Step Equations and Inequalities
- 8.1 Problem-Solving Strategy: Write an Equation to Solve Two-Step Problems R34
- 8.2 Simplifying and Solving R35
- 8.3 Comparing Equations and Inequalities R36
- 8.4 Solving Inequalities R37

CHAPTER 9 — Exploring Linear Equations
- 9.1 Graphing Ordered Pairs R38
- 9.2 Relations R39
- 9.3 Functions R40
- 9.4 Linear Equations R41

CHAPTER 10 — Congruence, Symmetry, and Transformations
- 10.1 Congruent Line Segments and Angles R42
- 10.2 Symmetry R43
- 10.3 Transformations R44
- 10.4 Transformations on the Coordinate Plane R45

CHAPTER 11: Constructing and Drawing

- 11.1 Constructing Congruent Angles and Line Segments R46
- 11.2 Constructing Parallel and Perpendicular Lines .. R47
- 11.3 Classifying and Comparing Triangles R48
- 11.4 Constructing Congruent Triangles R49

CHAPTER 12: Picturing and Modeling Solid Figures

- 12.1 Solid Figures R50
- 12.2 Problem-Solving Strategy: Finding Patterns in Polyhedrons R51
- 12.3 Nets for Solid Figures R52
- 12.4 Drawing Three-Dimensional Figures R53

CHAPTER 13: Changing Geometric Shapes

- 13.1 Tessellations R54
- 13.2 Geometric Iterations R55
- 13.3 Self-Similarity R56
- 13.4 Fractals R57

CHAPTER 14: Ratios and Rates

- 14.1 Problem-Solving Strategy: Drawing a Diagram to Show Ratios R58
- 14.2 Ratios and Rates R59
- 14.3 Rates in Tables and Graphs R60
- 14.4 Finding Golden Ratios R61

CHAPTER 15: Ratios, Proportions, and Percents

- 15.1 Changing Ratios to Percents R62
- 15.2 Finding a Percent of a Number R63
- 15.3 Finding What Percent One Number Is of Another R64
- 15.4 Finding a Number When the Percent Is Known R65

CHAPTER 16: Ratios, Proportions, and Similarity

- 16.1 Similar Figures and Scale Factors R66
- 16.2 Proportions and Similar Figures R67
- 16.3 Areas of Similar Figures R68
- 16.4 Volumes of Similar Figures R69

CHAPTER 17: Applications of Similar Figures

- 17.1 Drawing Similar Figures R70
- 17.2 Scale Drawings R71
- 17.3 Using Maps R72
- 17.4 Indirect Measurement R73
- 17.5 Golden Rectangles R74

CHAPTER 18: Growing and Shrinking Patterns

- 18.1 Triangular Arrays R75
- 18.2 Pascal's Triangle R76
- 18.3 Repeated Doubling and Halving R77
- 18.4 Exponents and Powers R78

CHAPTER 19: Number Patterns

- 19.1 Exploring Patterns in Decimals R79
- 19.2 Patterns in Rational Numbers R80
- 19.3 Patterns in Sequences R81
- 19.4 Patterns in Exponents R82

CHAPTER 20: Collecting Data—Sampling

- 20.1 Choosing a Sample R83
- 20.2 Bias in Samples R84
- 20.3 Writing Survey Questions R85
- 20.4 Organizing and Displaying Results R86

CHAPTER 21: Analyzing Data

- 21.1 How Do Your Data Shape Up? R87
- 21.2 Central Tendencies R88
- 21.3 Using Appropriate Graphs R89
- 21.4 Misleading Graphs R90

CHAPTER 22: Data and Probability

- 22.1 Tree Diagrams and Sample Spaces R91
- 22.2 Finding Probability R92
- 22.3 Problem-Solving Strategy: Make a List: Combinations and Probability R93
- 22.4 Finding Permutations and Probability R94

CHAPTER 23: Experiments with Probability

- 23.1 Experimental Probability R95
- 23.2 Problem-Solving Strategy: Acting It Out by Using Random Numbers R96
- 23.3 Designing a Simulation R97
- 23.4 Geometric Probability R98

CHAPTER 24: Measuring Length and Area

- 24.1 Measuring and Estimating Lengths R99
- 24.2 Networks R100
- 24.3 Pythagorean Property R101
- 24.4 Problem-Solving Strategy: Using a Formula to Find the Area R102
- 24.5 Area of a Trapezoid R103

CHAPTER 25: Surface Area and Volume

- 25.1 Surface Areas of Prisms and Pyramids R104
- 25.2 Surface Areas of Cylinders R105
- 25.3 Volumes of Prisms and Pyramids R106
- 25.4 Volumes of Cylinders and Cones R107

CHAPTER 26: Changing Dimensions

- 26.1 Changing Areas R108
- 26.2 Making Changes with Scaling R109
- 26.3 Problem-Solving Strategy: Making a Model: Volume and Surface Area R110
- 26.4 Volumes of Changing Cylinders R111

CHAPTER 27: Percent: Spending and Saving

- 27.1 Percent and Sales Tax R112
- 27.2 Percent and Discount R113
- 27.3 Percent and Markup R114
- 27.4 Earning Simple Interest R115
- 27.5 Problem-Solving Strategy: Making a Table to Find Interest R116

CHAPTER 28: Showing Relationships

- 28.1 Graphs and Pictures R117
- 28.2 Relationships in Graphs R118
- 28.3 Graphing Relationships R119
- 28.4 Using Scatterplots R120

Name _____

LESSON 1.1

Sets of Numbers

Here are some sets of numbers:
- 1, 2, 3, . . .   ← *counting numbers*
- 0, 1, 2, 3, . . .   ← *whole numbers*
- . . . ⁻3, ⁻2, ⁻1, 0, 1, 2, 3, . . .   ← *integers*

The fractions and decimals you are familiar with, such as 0.4, ⁻2.1, and $\frac{3}{2}$, are rational numbers. Each can be located on a number line.

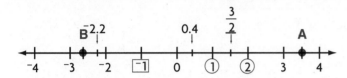

Use the number line above for Exercises 1–5. **Answers will vary. Possible answers are shown.**

1. Circle two counting numbers.
2. Draw a square around a negative integer.
3. Underline a rational number that is not an integer.
4. Draw a point to show the position of the number 3.5. Label it *A*.
5. Draw a point to show the position of the number $⁻2\frac{2}{3}$. Label it *B*.

For Exercises 6–8, match the description with its example.

6. a whole number   __c__
7. a number that is rational but not an integer   __a__
8. an integer but not a whole number   __b__

a. ⁻7.3
b. ⁻134
c. 0

9. On the number line, draw tick marks and label them with the numbers 0 to 12.

   0 ① 2 ③ 4 ⑤ 6 ⑦ 8 ⑨ 10 ⑪ 12

10. Circle the odd numbers on the number line above.
11. Underline the prime numbers on the number line above.
12. On the number line below, show the positions of 0.6, $\frac{⁻3}{4}$, and 0.2.

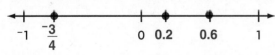

13. Show how you would use a number line to help you find some of the rational numbers between 6.6 and 6.7. **Answers will vary; check students' work.**

TAKE ANOTHER LOOK   R1

Understanding Rational Numbers

LESSON 1.2

A rational number can be written in many different ways.
The number $2\frac{1}{2}$ can be written as 2.5 or as $\frac{5}{2}$. These are all different names for the same number.

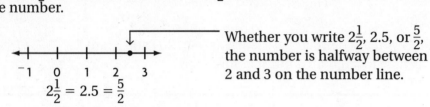

Whether you write $2\frac{1}{2}$, 2.5, or $\frac{5}{2}$, the number is halfway between 2 and 3 on the number line.

To graph a number on a number line, place a dot on the line.

Graph $^-1.25$ and $\frac{3}{5}$.

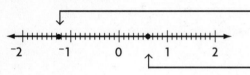

$^-1.25$ is $\frac{1}{4}$ of the way from $^-1$ to $^-2$.

To find $\frac{3}{5}$, divide the line between 0 and 1 into 5 equal parts.

Give at least three other names for each rational number. **Possible answers are given.**

1. 4 $\frac{4}{1}, \frac{8}{2}, 4.0$

2. $^-3\frac{1}{2}$ $^-3.5, ^-3.50, \frac{^-7}{2}$

3. $\frac{2}{3}$ $\frac{4}{6}, \frac{6}{9}, \frac{8}{12}$

4. $\frac{^-1}{8}$ $\frac{^-2}{16}, ^-0.125, \frac{^-4}{32}$

5. 1.6 $1.60, 1\frac{6}{10}, 1\frac{3}{5}$

6. $\frac{4}{10}$ $\frac{40}{100}, \frac{2}{5}, 0.4$

7. $^-5$ $\frac{^-5}{1}, ^-5.0, \frac{^-10}{2}$

8. 6.3 $6\frac{3}{10}, \frac{63}{10}, 6.30$

9. $^-4\frac{1}{4}$ $^-4.25, ^-4\frac{2}{8}, ^-4\frac{3}{12}$

10. 6 $6.0, \frac{6}{1}, \frac{12}{2}$

11. 0.18 $\frac{18}{100}, \frac{180}{1{,}000}, 0.180$

12. $^-2\frac{1}{3}$ $^-2\frac{4}{12}, ^-2\frac{10}{30}, ^-2\frac{2}{6}$

13. $1\frac{1}{8}$ $1.125, 1\frac{10}{80}, 1\frac{2}{16}$

14. 0.04 $0.040, \frac{4}{100}, \frac{1}{25}$

15. $^-1.8$ $^-1\frac{8}{10}, ^-1\frac{4}{5}, ^-1.80$

16. 3.2 $3.20, 3\frac{2}{10}, 3\frac{1}{5}$

17. $^-1$ $\frac{^-1}{1}, \frac{^-5}{5}, ^-1.0$

18. $7\frac{1}{2}$ $\frac{15}{2}, 7.5, 7\frac{10}{20}$

19. 0.45 $0.450, \frac{45}{100}, \frac{9}{20}$

20. $^-6.1$ $^-6.10, ^-6\frac{1}{10}, ^-6.100$

Graph each number on the number lines below.

21. 1
22. 1.5
23. $^-1.5$
24. $^-0.75$

25. $\frac{^-1}{8}$
26. 1.375
27. $^-1.5$
28. 0.625

R2 TAKE ANOTHER LOOK

Name _____

LESSON 1.3

Parts as Percents

Glass A holds 8 oz and is filled with 8 oz of juice. It is 100% filled.

Glass B has 4 oz of juice, so it is one-half filled. It is 50% filled.

Glass C has 2 oz of juice, so it is one-quarter filled. It is 25% filled.

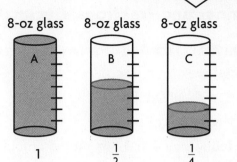

Write the percent to which each glass is filled.

1.    2. 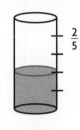   3. 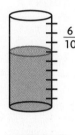   4.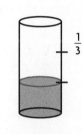

   75%   **40%**   **60%**   **about 33%**

Complete.

5. Percent means parts out of ___**100**___.

Shade each glass so that it appears filled to the indicated percent.

6. 75% 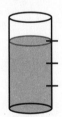   7. 30% 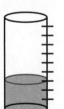   8. 60% 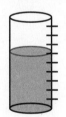   9. 90%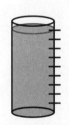

For each of the following, write the percent for the given amount of juice in a glass of the given size.

	Size of Glass	Amount in Glass	Percent Filled
10.	8 oz	6 oz	75%
11.	10 oz	2 oz	20%
12.	8 oz	1 oz	12.5%
13.	12 oz	3 oz	25%

	Size of Glass	Amount in Glass	Percent Filled
14.	8 oz	5 oz	62.5%
15.	10 oz	7 oz	70%
16.	6 oz	3 oz	50%
17.	12 oz	9 oz	75%

TAKE ANOTHER LOOK

Name _____

LESSON 1.4

Connecting Fractions, Decimals, and Percents

It can help to think of decimals, fractions, and percents as parts of a dollar.

25¢ = $0.25

25¢ = $\frac{1}{4}$ of a dollar

25¢ = 25% of a dollar

The following table shows money amounts written as decimals, percents, and fractions.

Money Amount	Decimal	Percent	Fraction
$0.05	0.05	5%	$\frac{5}{100} = \frac{1}{20}$
$0.17	0.17	17%	$\frac{17}{100}$
$0.25	0.25	25%	$\frac{25}{100} = \frac{1}{4}$
$1.50	1.50	150%	$\frac{150}{100} = 1\frac{1}{2}$

Write each of the following amounts of money as a percent of one dollar and as a fraction in simplest form.

1. $0.40   40%; $\frac{2}{5}$

2. $0.35   35%; $\frac{7}{20}$

3. $0.90   90%; $\frac{9}{10}$

4. $0.75   75%; $\frac{3}{4}$

5. $2.00   200%; $\frac{2}{1}$

6. $0.12   12%; $\frac{3}{25}$

7. $1.50   150%; $1\frac{1}{2}$

8. $0.01   1%; $\frac{1}{100}$

Consider the following fractions as parts of one dollar. Write each as an amount of money and as a percent.

9. $\frac{1}{2}$ $0.50; 50%

10. $\frac{7}{10}$ $0.70; 70%

11. $\frac{45}{100}$ $0.45; 45%

12. $\frac{3}{20}$ $0.15; 15%

13. $\frac{5}{2}$ $2.50; 250%

14. $\frac{3}{25}$ $0.12; 12%

15. $\frac{3}{50}$ $0.06; 6%

16. $\frac{1}{5}$ $0.20; 20%

Write each percent as an amount of money and as a fraction in simplest form.

17. 60% $0.60; $\frac{3}{5}$

18. 63% $0.63; $\frac{63}{100}$

19. 72% $0.72; $\frac{18}{25}$

20. 100% $1; $\frac{1}{1}$

R4   TAKE ANOTHER LOOK

Making Circle Graphs

LESSON 1.5

A circle graph shows data as parts of a whole. Remember, the sum of the central angles of a circle equals 360°.

Make a circle graph to show the following data from a class survey about favorite hobbies: sports, 40%; crafts, 15%; shopping, 10%; reading, 20%; computer games, 15%.

Step 1
Find the measure of the central angle that will represent each part of the survey.

40%: 0.40 × 360° = 144°
10%: 0.10 × 360° = 36°
15%: 0.15 × 360° = 54°
15%: 0.15 × 360° = 54°
20%: 0.20 × 360° = 72°

Step 2
Draw a circle. Draw one radius to give you a starting point for drawing your central angles.

Step 3
Use a protractor to draw each central angle.

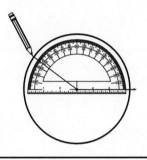

Step 4
Label each section of your graph. Give your graph a title.

FAVORITE HOBBIES

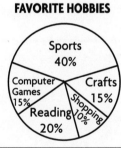

Find the central angle measure for each percent of a circle.

1. 60% ___216°___
2. 25% ___90°___
3. 45% ___162°___
4. 5% ___18°___
5. 80% ___288°___
6. 55% ___198°___
7. 50% ___180°___
8. 12.5% ___45°___

9. Make a circle graph for a company with the following percent of sales for each division: West, 20%; Central, 35%; South, 30%; North, 15%. **Check students' graphs. Measures: West, 72°; Central, 126°; South, 108°; North, 54°**

TAKE ANOTHER LOOK R5

Name _____

LESSON 2.1

Using Exponents

Can you find the fifth number in this pattern? the tenth?

$$5, 25, 125, 625, \ldots$$

The pattern uses multiplication, so the numbers get larger fast. The fifth number is 3,125, and the tenth number is almost 10 million!

Look at the pattern written more simply.

$$5^1, 5^2, 5^3, 5^4, \ldots$$

The expression 5^4 means $5 \times 5 \times 5 \times 5$ and is read *5 to the fourth power*. The base is 5 and the exponent is 4. Use your calculator to check the following calculations.

$5^4 = 5 \times 5 \times 5 \times 5 = 625$
$5^5 = 5 \times 5 \times 5 \times 5 \times 5 = 3{,}125$
$5^{10} = 5 \times 5 \times 5 \times 5 \times 5 \times 5 \times 5 \times 5 \times 5 \times 5 = 9{,}765{,}625$

Scientific notation is another way to write numbers more simply. It shows very large numbers in an easier way.

Remember, the exponent tells you how many places to move the decimal point.

$$4.3 \times 10^4 = 43000 = 43{,}000 \leftarrow \text{Move the decimal point 4 places.}$$

Find the value.

1. 2^3 __8__
2. 3^2 __9__
3. 5^4 __625__
4. 4^5 __1024__

5. 10^3 __1,000__
6. 1^5 __1__
7. 7^2 __49__
8. 20^3 __8,000__

9. 2^8 __256__
10. 10^8 __100,000,000__
11. 6^3 __216__
12. 9^4 __6,561__

13. 2^6 __64__
14. 8^5 __32,768__
15. 30^2 __900__
16. 50^2 __2,500__

Multiply so that each of the following is written in standard form.

17. 2.3×10^4 __23,000__
18. 1.74×10^5 __174,000__
19. 6.04×10^7 __60,400,000__
20. 3.0×10^{10} __30,000,000,000__

R6 TAKE ANOTHER LOOK

Name _____

LESSON 2.2

Exploring Decimal and Binary Numbers

Each power of 2 below represents a bag of pennies. A "1" below a power of 2 means you take that bag of pennies. A "0" means you do not take that bag of pennies.

Power of 2	2^7	2^6	2^5	2^4	2^3	2^2	2^1	2^0
Number of pennies in bag	128	64	32	16	8	4	2	1
Take it?	0	0	0	1	0	1	0	1

Which bags does 10101 tell you to take? The 2^4 bag, the 2^2 bag, and the 2^0 bag.

How many pennies will you have? Add the number of pennies in each bag.

$$16 + 4 + 1 = 21$$

So, the binary number 10101_{two} has a decimal value of 21.

Given each binary number, which bags of pennies will you take? How many pennies will you have?

1. 11_{two}

    2^1 and 2^0; 3 pennies

2. 101_{two}

    2^2 and 2^0; 5 pennies

3. 10000000_{two}

    2^7; 128 pennies

4. 1101_{two}

    2^3, 2^2, and 2^0; 13 pennies

5. 11011011_{two}

    2^7, 2^6, 2^4, 2^3, 2^1, and 2^0; 219 pennies

6. 1110111_{two}

    2^6, 2^5, 2^4, 2^2, 2^1, and 2^0; 119 pennies

7. 11011_{two}

    2^4, 2^3, 2^1, and 2^0; 27 pennies

8. 10111101_{two}

    2^7, 2^5, 2^4, 2^3, 2^2, and 2^0; 189 pennies

9. 1010111_{two}

    2^6, 2^4, 2^2, 2^1, and 2^0; 87 pennies

10. 10111_{two}

    2^4, 2^2, 2^1, and 2^0; 23 pennies

11. Using the binary number system, you can write any number using only

    the digits __0__ and __1__.

TAKE ANOTHER LOOK

Name _____

LESSON 2.3

Modeling Squares and Square Roots

To estimate the value of a square root, it is helpful to know the squares of the integers between 1 and 15.

Between which two integers is $\sqrt{90}$?
- $9^2 = 81$ and $10^2 = 100$
- 90 is between 81 and 100.
- $\sqrt{90}$ is between $\sqrt{81}$ and $\sqrt{100}$.
- So, $\sqrt{90}$ is between 9 and 10.

Complete the table.

1.

Integer	1	2	3	4	5	6	7	8	9	10	11	12	13	14	15
Square of Integer	1	4	9	16	25	36	49	64	81	100	121	144	169	196	225

Using the table from Exercise 1, find two integers between which the square root falls.

2. $\sqrt{120}$ __10 and 11__ 3. $\sqrt{71}$ __8 and 9__ 4. $\sqrt{224}$ __14 and 15__

5. $\sqrt{55}$ __7 and 8__ 6. $\sqrt{175}$ __13 and 14__ 7. $\sqrt{143}$ __11 and 12__

8. $\sqrt{17}$ __4 and 5__ 9. $\sqrt{128}$ __11 and 12__ 10. $\sqrt{95}$ __9 and 10__

Solve.

11. A square lawn has an area of 220 ft². A fence for one side of the lawn will measure between what two lengths?

    _____14 ft and 15 ft_____

12. A square piece of paper has an area of 186 cm². A side of the paper measures between what two lengths?

    _____13 cm and 14 cm_____

13. A square photo has an area of 80 in.². To the nearest inch, what is the length of one side of the photo?

    _____9 in._____

14. A square farm measures 48 mi². To the nearest mile, what is the measure of one side?

    _____7 mi_____

R8 TAKE ANOTHER LOOK

Name _____

LESSON 2.4

Problem-Solving Strategy

Using Guess and Check to Find Square Roots

The strategy *guess and check* can be used to find a square root. It helps to use an *averaging method* as you make each guess.

Your dog needs a play area of 1,000 ft^2. If the play area is a square, what should it measure on each side, to the nearest foot?

Suppose your first guess is 20 ft.

Step 1 Divide. $1,000 \div 20 = 50$
$\sqrt{1,000}$ is between 20 and 50.

Step 2 Find the average of your guess and the new quotient.
$20 + 50 = 70 \quad 70 \div 2 = 35$

Step 3 Divide 1,000 by the average. $1,000 \div 35 = 28.6$, or 29
$\sqrt{1,000}$ is between 29 and 35.

Step 4 Find the average of the last two quotients.
$35 + 29 = 64 \quad 64 \div 2 = 32$

Step 5 Divide 1,000 by the average. $1,000 \div 32 = 31.25$
$\sqrt{1,000}$ is between 31.25 and 32.

You have a good estimate when the quotient is close to the divisor.

For each of the following, find the square root using the strategy guess and check, with averaging, to improve your guess.

1. A square plot of land measures 800 m^2. Find the length of a side to the nearest meter. __**28 m**__

2. A square computer table top measures 12.25 ft^2. Find the length of a side. __**3.5 ft**__

3. Colorado has an area of 104,100 mi^2. If Colorado were a square, what would be the length of a side, to the nearest 10 miles? __**320 mi**__

4. Carmen and Oran had an argument about the square root of 460. Carmen said it was closer to 21. Oran said it was closer to 22. Who was right? __**Carmen**__

5. A bag of grass seed is intended to cover 500 ft^2. If a square lawn can be seeded by 10 bags, what is the measure of one side of the lawn, to the nearest foot? __**71 ft**__

6. A square garden has an area of 512 ft^2. Find the length of a side to the nearest foot. __**23 ft**__

TAKE ANOTHER LOOK

Name _____

LESSON 2.5

Repeated Calculations

You can form a geometric pattern using an iteration. An iteration process repeats something over and over again.

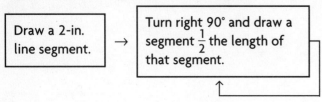

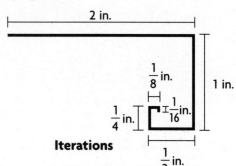

Iterations

Continue the process four more times.

For Problems 1–3, use the iteration process shown.

1. Start at the center of an $8\frac{1}{2}$-in. × 11-in. sheet of paper.

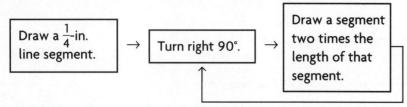

Continue the process as far as you can. **Check students' drawings.**

2. Start at the center of an $8\frac{1}{2}$-in. × 11-in. sheet of paper.

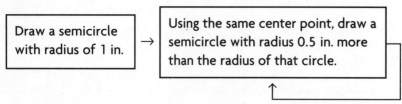

Continue the process as far as you can. What is the radius of the fifth semicircle? __3 in. Check students' drawings.__

3. Start at the bottom of an $8\frac{1}{2}$-in. × 11-in. sheet of paper.

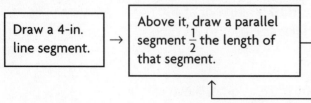

Continue the process as far as you can. How long is the fifth line segment? __$\frac{1}{4}$ in. Check students' drawings.__

TAKE ANOTHER LOOK

Name _____

LESSON 3.1

Estimating Sums and Differences

You can round to estimate sums and differences. For example, seven people weighing 113 lb, 280 lb, 135 lb, 302 lb, 178 lb, 94 lb, and 156 lb enter an elevator. The weight limit of the elevator is 1,200 lb. Estimate whether the elevator has reached its weight limit.

Step 1 Round each weight with the nearest ten pounds to find the sum more easily.

113 + 280 + 135 + 302 + 178 + 94 + 156
↓   ↓   ↓   ↓   ↓   ↓   ↓
110 + 280 + 140 + 300 + 180 + 90 + 160 = 1,260

Step 2 Compare the estimate with the weight limit of the elevator.

1,260 > 1,200

The elevator is filled beyond its weight limit.

---

1. The same elevator carries six passengers weighing 225 lb, 186 lb, 319 lb, 103 lb, 212 lb, 58 lb. Estimate whether the elevator has reached its weight limit.

   Step 1 Round each weight to the nearest ten pounds to find the sum.

   225 + 186 + 319 + 103 + 212 + 58
   ↓   ↓   ↓   ↓   ↓   ↓
   230 + 190 + 320 + 100 + 210 + 60 = 1,110

   Step 2 Compare the estimate with the weight limit of the elevator.

   1,110 < 1,200

   The elevator __is not__ filled beyond its weight limit.

George has $35.00 to buy school supplies. In Exercises 2 and 3, the prices of the supplies are given. Complete to estimate whether George has enough money to buy the supplies.

2. $15.68, $2.75, $3.25, $1.79, and $5.39

   Step 1 Round to the nearest dollar to find the sum.

   15.68 + 2.75 + 3.25 + 1.79 + 5.39
   ↓   ↓   ↓   ↓   ↓
   16 + 3 + 3 + 2 + 5 = 29

   Step 2 Compare the estimate with the amount of money George can spend.

   $29.00 < $35.00

   George __has__ enough money to buy the supplies.

3. $3.79, $11.29, $7.99, $4.95, and $12.19 $40.00 > $35.00

   George __does not have__ enough money to buy the supplies.

TAKE ANOTHER LOOK   R11

Name _____

LESSON 3.2

Multiplying Whole Numbers and Decimals

You need to know the number of decimal places in each factor when multiplying decimals.

        0.43   ←   two decimal places
        14.1   ←   one decimal place
        0.906   ←   three decimal places
        6.0001 ←   four decimal places
        20   ←   zero decimal places

Multiply 1.5 by 0.005.

Step 1 Multiply as whole numbers.
 	
    1.5
 ×0.005
     75
	Step 2 Identify the number of decimal places in each factor. Find the total number of decimal places.
	
	    1.5 1 decimal place
	×0.005 + 3 decimal places
	     75 4 decimal places

Step 3 Count 4 decimal places from right to left in the product.
 	
    1.5
 ×0.005
   .75
	Step 4 Write zeros as placeholders.
	
	    1.5
	 ×0.005
	 0.0075

Complete to multiply.

1.   0.7   1 decimal place
   ×0.3   +1 decimal place
  0.21   **2** decimal place(s)

2.   0.2   **1** decimal place(s)
  × 5   + **0** decimal place(s)
  1.0   **1** decimal place(s)

3.   **2** decimal place(s)
    0.43   + **1** decimal place(s)
  × 0.6   **3** decimal place(s)
 0.258

4.   **3** decimal place(s)
    8.012   + **2** decimal place(s)
  × 0.08   **5** decimal place(s)
 0.64096

5.   **3** decimal place(s)
   0.012   + **1** decimal place(s)
  × 3.2   **4** decimal place(s)
 0.0384

6.   **2** decimal place(s)
    0.34   + **2** decimal place(s)
  ×0.21   **4** decimal place(s)
 0.0714

7.   **4** decimal place(s)
   0.0073 + **2** decimal place(s)
  × 0.16   **6** decimal place(s)
 0.001168

8.   **3** decimal place(s)
   1.045   + **3** decimal place(s)
  ×0.001   **6** decimal place(s)
 0.001045

R12   TAKE ANOTHER LOOK

Name _____

LESSON 3.3

Dividing Whole Numbers and Decimals

When you divide any numbers, the number of decimal places in the divisor tells you where the decimal point goes in the quotient.

When you divide by a whole number, here is how you place the decimal point.

Solve $0.35 \div 5$.

- Find the decimal point in the dividend.

    $5\overline{)0.35}$

- Place the decimal point in the quotient directly above the decimal point in the dividend.

    $5\overline{)0.35}$

- Every decimal place to the right of the decimal point in the dividend must have a number above it.

    $.0$
    $5\overline{)0.35}$

- Divide.

    $.07$
    $5\overline{)0.35}$

When you divide by a decimal, write an equivalent division problem so the divisor is a whole number.

Solve $5 \div 0.02$.

- Find the number of decimal places in the divisor. There are 2 decimal places.

    $0.02\overline{)5}$

- Write zeros so you have the same number of decimal places in the dividend as in the divisor.

    $0.02\overline{)5.00}$

- Multiply *both* divisor and dividend by 10^2, or 100. Move both decimal points two places to the right.

    $0.02\overline{)5.00}$

- Place the decimal point in the quotient.

    $2\overline{)500}$

- Divide.

    $250.$
    $2\overline{)500}$

Complete to solve.

1. $15.15 \div 0.15$

    How many decimal places are in the divisor? __2__

    How many decimal places to the right do you move *both* decimal points? __2__

    What is the quotient? __101__

2. $0.28 \div 4$

    How many decimal places are in the divisor? __0__

    How many decimal places to the right do you move *both* decimal points? __0__

    What is the quotient? __0.07__

TAKE ANOTHER LOOK R13

Name _____

LESSON 3.4

Order of Operations

In sports, there are certain rules to help avoid confusion. To avoid confusion in mathematics and tell which value is correct, operations are done in a special order.

Find the value of $18 \div 3^2 \times (12 - 4) - 7 + 4$. Underline each step as you do it.

$18 \div 3^2 \times \underline{(12 - 4)} - 7 + 4$ • Do operations inside parentheses.

$18 \div \underline{3^2} \times 8 - 7 + 4$ • Clear the exponents.

$\underline{18 \div 9} \times 8 - 7 + 4$ • Multiply and divide from left to right.

$\underline{2 \times 8} - 7 + 4$

$\underline{16 - 7} + 4$ • Add and subtract from left to right.

$9 + 4$

13

Give the correct order of operations to find the value.

1. $6 \div 3 + 4^2 - (5 + 4) = 6 \div 3 + 4^2 - \boxed{9}$ Operate inside parentheses.

$= 6 \div 3 + \boxed{16} - \boxed{9}$ Clear the exponent.

$= \boxed{2} + \boxed{16} - \boxed{9}$ Divide.

$= \boxed{18} - \boxed{9}$ Add.

$= \boxed{9}$ _____ Subtract.

2. $12 + 20 \div 4 - 5 = 12 + \boxed{5} - 5$ Divide.

$= \boxed{17} - 5$ _____ Add.

$= \boxed{12}$ _____ Subtract.

3. $3^2 + (9 + 15) \div (8 - 5) + 6 = 3^2 + \boxed{24} \div \boxed{3} + 6$ _____ Operate inside

_____ parentheses.

$= \boxed{9} + \boxed{24} \div \boxed{3} + 6$ _____ Clear the exponent.

$= \boxed{9} + \boxed{8} + 6$ _____ Divide.

$= \boxed{17} + 6$ _____ Add.

$= \boxed{23}$ _____ Add.

R14 TAKE ANOTHER LOOK

Adding and Subtracting Fractions

Lesson 4.1

When fractions have the same denominator, you add or subtract only the numerators.

When fractions have different denominators, you must rewrite them as equivalent fractions with the same denominator.

Add $\frac{5}{6}$ ft, $\frac{3}{10}$ ft, and $\frac{4}{15}$ ft.

Step 1 Find the *least common denominator*, or LCD, for the denominators.
• Write the multiples of 6, 10, and 15.
   6, 12, 18, 24, <u>30</u>, . . .
   10, 20, <u>30</u>, . . .
   15, <u>30</u>, . . .
• Find the smallest common multiple.
• The LCD is 30.
	Step 2 Write equivalent fractions using the LCD.
	$\frac{5 \times 5}{6 \times 5} = \frac{25}{30}$
	$\frac{3 \times 3}{10 \times 3} = \frac{9}{30}$
	$\frac{4 \times 2}{15 \times 2} = \frac{8}{30}$

Step 3 Add the fractions. • Add the numerators. • *Never* add the denominators. $\frac{25}{30} + \frac{9}{30} + \frac{8}{30} = \frac{25 + 9 + 8}{30}$ $= \frac{42}{30}$	**Step 4** Write the answer in simplest form. • Divide numerator and denominator by any common factors. $\frac{42 \div 6}{30 \div 6} = \frac{7}{5}$ • If numerator > denominator, change to a mixed number. $\frac{7}{5} = 1\frac{2}{5}$

Complete to add or subtract.

1. $\frac{2}{7} + \frac{1}{3}$

   Step 1 Find the LCD for 7 and 3. Write the multiples of 7 and 3.

   7, __14__, __21__, __28__, __35__, __42__, __49__, . . .

   3, __6__, __9__, __12__, __15__, __18__, __21__, . . .

   The LCD is __21__.

   Step 2 Write equivalent fractions using the LCD.

   $\frac{2 \times \boxed{3}}{7 \times \boxed{3}} = \frac{6}{21}$ $\frac{1 \times \boxed{7}}{3 \times \boxed{7}} = \frac{7}{21}$

   Step 3 Add the numerators. Keep the same denominator.

   $\frac{\boxed{6} + \boxed{7}}{\boxed{21}} = \frac{\boxed{13}}{\boxed{21}}$

   Step 4 Write the answer in simplest form.

   $\frac{13}{21}$

2. $\frac{7}{10} - \frac{3}{5}$ LCD = __10__ $\frac{7}{10} - \frac{3}{5} = \frac{\boxed{7}}{\boxed{10}} - \frac{\boxed{6}}{\boxed{10}} = \frac{\boxed{1}}{\boxed{10}}$

Name _____

LESSON 4.2

Adding and Subtracting Mixed Numbers

Subtract. $3\frac{1}{2} - 1\frac{5}{8}$

Step 1 Find the LCD.
• Write the multiples of 2 and 8.
  2, 4, 6 _8_, 10, . . .
  _8_, 16, 24, . . .
• The LCD is 8.
	Step 2 Write equivalent fractions using the LCD.
	$3\frac{1}{2} = 3\frac{1\times 4}{2\times 4} = 3\frac{4}{8}$
	$1\frac{5}{8} = 1\frac{5}{8}$

Step 3 Subtract the fraction part of the mixed numbers. $3\frac{4}{8} = 2\frac{12}{8}$ $-1\frac{5}{8} = -1\frac{5}{8}$ $\phantom{-1\frac{5}{8} = -1}\frac{7}{8}$ • Since 5 > 4, rename $3\frac{4}{8}$. $3\frac{4}{8} = 2 + \frac{8}{8} + \frac{4}{8} = 2\frac{12}{8}$	**Step 4** Subtract the whole-number part of the mixed numbers. $2\frac{12}{8}$ $-0\frac{5}{8}$ $0\frac{7}{8}$ • Check to see that your answer is in simplest form.

Complete to subtract. $5\frac{1}{5} - 3\frac{5}{6}$

1. Step 1 Find the LCD.
- Write the multiples of 5 and 6.

  5, __10__, __15__, __20__, __25__, __30__, __35__, . . .

  6, __12__, __18__, __24__, __30__, __36__, __42__, . . .

- The LCD is __30__.

2. Step 2 Write equivalent fractions using the LCD.

$5\frac{1}{5} = 5\dfrac{1 \times \boxed{6}}{5 \times \boxed{6}} = 5\dfrac{6}{30}$

$3\frac{5}{6} = 3\dfrac{5 \times \boxed{5}}{6 \times \boxed{5}} = 3\dfrac{25}{30}$

3. Step 3 Subtract the fraction part of the mixed numbers.
- Rename $5\frac{6}{30}$.

$5\dfrac{6}{30} = 4 + \dfrac{\boxed{30}}{30} + \dfrac{\boxed{6}}{30} = 4\dfrac{\boxed{36}}{30}$

$4\dfrac{\boxed{36}}{30}$
$-3\dfrac{\boxed{25}}{30}$
$\dfrac{\boxed{11}}{30}$

4. Step 4 Subtract the whole number part of the mixed numbers.
- Check that your answer is in simplest form.

$4\dfrac{\boxed{36}}{30}$
$-3\dfrac{\boxed{25}}{30}$
$1\dfrac{\boxed{11}}{30}$

Name _____

LESSON 4.3

Estimating Sums and Differences

The Caron family want to paint their living room. One gallon of paint covers 400 ft². The Carons do not need an *exact* measure of the walls to buy the correct amount of paint. They can use an estimate. When you estimate with mixed numbers, you can round the fractions to 0, $\frac{1}{2}$, or 1.

Estimate $\frac{1}{10} + 2\frac{2}{3}$.

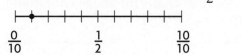
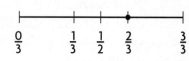

Step 1 Is $\frac{1}{10}$ closest to 0, $\frac{1}{2}$, or 1 on a number line?
Since $\frac{1}{10}$ is closest to 0, round $\frac{1}{10}$ to 0.

Step 2 Since $2\frac{2}{3}$ is a mixed number, round just the fraction.
Is $\frac{2}{3}$ closest to 0, $\frac{1}{2}$, or 1 on a number line?
Since $\frac{2}{3}$ is closest to $\frac{1}{2}$, round $\frac{2}{3}$ to $\frac{1}{2}$.
To round the mixed number, add. $2 + \frac{1}{2} = 2\frac{1}{2}$

Step 3 Add the estimates. $0 + 2\frac{1}{2} = 2\frac{1}{2}$

So, the sum is about $2\frac{1}{2}$.

Show each fraction on the number line. Complete to estimate $4\frac{3}{8} + 2\frac{7}{9}$.

1.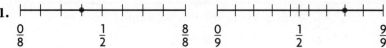

2. Round $4\frac{3}{8}$.

   The whole-number part of $4\frac{3}{8}$ is __4__. The fraction part of $4\frac{3}{8}$ is $\frac{3}{8}$.

   Round the fraction part to $\frac{1}{2}$.

   Round $4\frac{3}{8}$ to __4__ + $\frac{1}{2}$, or $4\frac{1}{2}$.

3. Round $2\frac{7}{9}$.

   The whole-number part of $2\frac{7}{9}$ is __2__. The fraction part of $2\frac{7}{9}$ is $\frac{7}{9}$.

   Round the fraction part to __1__.

   Round $2\frac{7}{9}$ to 2 + __1__, or __3__.

4. Add the estimates. $4\frac{1}{2}$ + __3__ = $7\frac{1}{2}$

Name _____

LESSON 4.4

Multiplying and Dividing Fractions and Mixed Numbers

You learned how to divide fractions and mixed numbers using multiplication and a reciprocal. Another way is to use common denominators. To divide fractions that have a common denominator, divide the numerators.

Marta buys $3\frac{3}{4}$ lb of pecans. If she divides the pecans into bags containing $\frac{7}{8}$ lb each, how many bags does she fill?

Step 1 Write any mixed numbers as fractions. $3\frac{3}{4} \div \frac{7}{8} = \frac{15}{4} \div \frac{7}{8}$	**Step 2** Write the fractions using a common denominator. $\frac{15}{4} \div \frac{7}{8} = \frac{30}{8} \div \frac{7}{8}$
Step 3 Divide the numerators. $30 \div 7 = 4\frac{2}{7}$	**Step 4** Check that your answer is simplified. $4\frac{2}{7}$ ✔ simplified

Marta fills $4\frac{2}{7}$ bags.

---

1. Complete to divide. $6\frac{1}{8} \div 1\frac{1}{4}$

Step 1 Write any mixed numbers as fractions. $6\frac{1}{8} \div 1\frac{1}{4} = \frac{49}{8} \div \frac{5}{4}$	**Step 2** Write the fractions using a common denominator. $\frac{49}{8} \div \frac{5}{4} = \frac{49}{8} \div \frac{10}{8}$
Step 3 Divide the numerators. $49 \div 10 = 4\frac{9}{10}$	**Step 4** Write the simplified answer. $4\frac{9}{10}$

Divide using common denominators. Write the answer in simplest form.

2. $\frac{14}{15} \div \frac{7}{15}$ __2__
3. $3\frac{1}{4} \div \frac{2}{3}$ __$4\frac{7}{8}$__
4. $\frac{3}{7} \div \frac{9}{14}$ __$\frac{2}{3}$__
5. $\frac{7}{8} \div \frac{1}{8}$ __7__

6. $1\frac{1}{5} \div \frac{1}{4}$ __$4\frac{4}{5}$__
7. $\frac{1}{2} \div \frac{3}{10}$ __$1\frac{2}{3}$__
8. $3\frac{3}{4} \div \frac{3}{8}$ __10__
9. $\frac{3}{4} \div \frac{3}{8}$ __2__

10. $4\frac{5}{6} \div \frac{1}{2}$ __$9\frac{2}{3}$__
11. $3\frac{1}{2} \div 2\frac{1}{4}$ __$1\frac{5}{9}$__
12. $1\frac{9}{10} \div \frac{3}{10}$ __$6\frac{1}{3}$__
13. $\frac{5}{6} \div \frac{1}{30}$ __25__

14. $7\frac{1}{2} \div 1\frac{3}{4}$ __$4\frac{2}{7}$__
15. $1\frac{3}{4} \div \frac{1}{2}$ __$3\frac{1}{2}$__
16. $1\frac{2}{3} \div \frac{4}{5}$ __$2\frac{1}{12}$__
17. $3\frac{1}{5} \div \frac{4}{5}$ __4__

18. $4\frac{1}{2} \div \frac{5}{6}$ __$5\frac{2}{5}$__
19. $4\frac{1}{4} \div \frac{3}{4}$ __$5\frac{2}{3}$__
20. $2\frac{1}{2} \div 1\frac{3}{5}$ __$1\frac{9}{16}$__
21. $6\frac{1}{2} \div \frac{3}{4}$ __$8\frac{2}{3}$__

R18 **TAKE ANOTHER LOOK**

Name _____

LESSON 4.5

Problem-Solving Strategy

Solve a Simpler Problem

Large numbers make a problem more difficult to solve. You can use a simpler problem related to the difficult one to help you solve.

Ryan has 5,000 baseball cards for sale in his shop. He sells $\frac{1}{10}$ of them to his best customers. He sells $\frac{1}{2}$ of the remaining cards in a special sale. How many cards does Ryan have left?

Use smaller numbers to solve a related simpler problem.
Divide 5,000 by 100 to let 50 represent the number of cards with which Ryan started.

- Ryan sells $\frac{1}{10}$ of the cards. $\frac{1}{10} \times 50 = 5$
- Find the remaining cards. $50 - 5 = 45$
- Find the number of cards sold in the special sale. $\frac{1}{2} \times 45 = 22.5$
- Multiply by 100 to get the actual number of cards left. $22.5 \times 100 = 2,250$

So, Ryan has 2,250 cards left.

Complete to solve.

Hideko's class collects 2,400 cans to recycle. The class cleans $\frac{1}{4}$ of the cans. Of the remaining cans, $\frac{1}{9}$ are damaged and cannot be recycled. How many of the remaining cans are acceptable for recycling?

- Use a smaller number to solve a simpler problem.

- Divide 2,400 by 100 to let __24__ represent the number of cans the class collects.

- The class cleans $\frac{1}{4}$ of the cans. How many cans are clean?

  $\frac{1}{4} \times$ __24__ = __6__

- How many cans still need to be cleaned? __24__ − __6__ = __18__
- Of these cans, $\frac{1}{9}$ are damaged. How many cans are damaged?

  $\frac{1}{9} \times$ __18__ = __2__

- How many of the remaining cans are acceptable for recycling?

  __18__ − __2__ = __16__

- Multiply this amount by 100 to find the actual number of remaining cans acceptable for recycling.

  __16__ × __100__ = __1,600__ ; __1,600__ cans

TAKE ANOTHER LOOK   R19

Name _____

LESSON
5.1

Adding Integers

This balance scale "weighs" positive and negative numbers. Negative numbers go on the left of the balance and positive numbers go on the right.

Find $^-11 + 8$.
The scale will tip to the left side, because it is $^-3$ "heavier."

Find $^-2 + 7$.
The scale will tip to the right side, because it is $^+5$ "heavier."

Find $^-1 + {}^-3$.
Both $^-1$ and $^-3$ go on the left side. The scale will tip to the left side, because it is $^-4$ "heavier."

Find how much "heavier" the lower side is.

1.
   5

2.
   $^-4$

3.
   1

4.
   0

5.
   $^+5$

6.
   $^-14$

Solve.

7. $7 + {}^-3$
   4

8. $^-2 + {}^-3$
   $^-5$

9. $^-9 + 4$
   $^-5$

Add the integers.

10. $^-3 + {}^-1$
    $^-4$

11. $^-7 + 9$
    2

12. $4 + {}^-9$
    $^-5$

13. $^-16 + 28$
    12

14. $11 + {}^-11$
    0

15. $^-25 + 11$
    $^-14$

16. $^-15 + {}^-21$
    $^-36$

17. $23 + {}^-19$
    4

18. $^-11 + 15$
    4

R20 TAKE ANOTHER LOOK

Name _____

LESSON 5.2

Subtracting Integers

The total value of the three cards shown is ⁻6.

- If the card with ⁺3 is taken away, the value of the cards decreases by 3.
- If the card with ⁻4 is taken away, the value of the cards increases by 4.

Suppose you have the cards shown. The total value of your cards is 11. What happens if you take away the ⁻8 card?

Cards 12 and 7 are left. The new value is 19.   11 − ⁻8 = 19

For Exercises 1–3, suppose you have the cards shown.

1. The total value of your cards is ⁻10.

   a. What is the value if you take away the 3 card? __⁻13__ So, ⁻10 − 3 = __⁻13__.

   b. What is the value if you take away the ⁻6 card? __⁻4__ So, ⁻10 − ⁻6 = __⁻4__.

   c. What is the value if you take away the ⁻7 card? __⁻3__ So, ⁻10 − ⁻7 = __⁻3__.

2. The total value of your cards is 12.

   a. What is the value if you take away the 7 card? __5__ So, 12 − 7 = __5__.

   b. What is the value if you take away the 13 card? __⁻1__ So, 12 − 13 = __⁻1__.

   c. What is the value if you take away the ⁻8 card? __20__ So, 12 − ⁻8 = __20__.

3. The total value of your cards is ⁻18.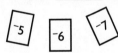

   a. What is the value if you take away the ⁻5 card? __⁻13__ So, ⁻18 − ⁻5 = __⁻13__.

   b. What is the value if you take away the ⁻6 card? __⁻12__ So, ⁻18 − ⁻6 = __⁻12__.

   c. What is the value if you take away the ⁻7 card? __⁻11__ So, ⁻18 − ⁻7 = __⁻11__.

Find the difference.

4. ⁻4 − ⁻2   5. 21 − 13   6. ⁻9 − 17   7. 31 − ⁻9   8. 15 − 18
   __⁻2__   __8__   __⁻26__   __40__   __⁻3__

9. ⁻7 − ⁻9   10. ⁻8 − ⁻8   11. 29 − ⁻2   12. 13 − 18   13. ⁻11 − 3
   __2__   __0__   __31__   __⁻5__   __⁻14__

TAKE ANOTHER LOOK   R21

Name _____

LESSON 5.3

Multiplying and Dividing Integers

The students at Wingdale Middle School are washing cars to earn money for books for the school library.

- Each car washed adds $4 to the treasury.
- Each book purchased subtracts $6 from the treasury.
- Twenty cars washed: $20 \times 4 = 80$ *Add $80 to the treasury.*
- Twelve books purchased: $12 \times {}^-6 = {}^-72$ *Subtract $72 from the treasury.*

Write a multiplication expression describing the following.

1. 25 cars washed 25×4
2. 7 books purchased $7 \times {}^-6$
3. 13 cars washed 13×4
4. 12 books purchased $12 \times {}^-6$

How much is added to or subtracted from the treasury? Use an equation to show your work.

5. 10 cars washed
   Adds $40; $10 \times 4 = 40$
6. 6 books purchased
   Subtracts $36; $6 \times {}^-6 = {}^-36$
7. 55 cars washed
   Adds $220; $55 \times 4 = 220$
8. 17 books purchased
   Subtracts $102; $17 \times {}^-6 = {}^-102$

For a summer job, Doug and Marissa are doing pool maintenance in the neighborhood. This week, they are draining and refilling the pools.

9. Will the amount of water being drained be represented by a positive or a negative number?

   negative number

10. Will the amount of water being put back be represented by a positive or a negative number?

    positive number

Doug and Marissa drain a 960-gal pool in 30 min. What is the average change in the amount of water in the pool, in gallons per minute?

${}^-960 \div 30 = {}^-32$ The average change is ${}^-32$ gal per minute.

Write a division equation to find the average change in the amount of water.

11. 1,200-gal pool drained in 40 min
    ${}^-1{,}200 \div 40 = {}^-30$ **gal per min**
12. 1,500-gal pool refilled in 50 min
    $1{,}500 \div 50 = 30$ **gal per min**
13. 1,000-gal pool refilled in 25 min
    $1{,}000 \div 25 = 40$ **gal per min**
14. 900-gal pool drained in 60 min
    ${}^-900 \div 60 = {}^-15$ **gal per min**

R22   TAKE ANOTHER LOOK

Adding and Subtracting Rational Numbers

LESSON 5.4

When adding rational numbers, you can use a number line to find the answer, just as you do for integers. Remember:

- A positive number is represented by an arrow to the right.
- A negative number is represented by an arrow to the left.

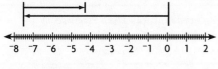

$^{-}7.5 + 3.2 = {^-}4.3$

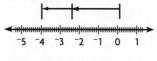

$^{-}2.5 + {^-}1.6 = {^-}4.1$

You know that subtracting an integer is the same as adding its opposite. The same is true for rational numbers. Make sure you change subtraction problems into addition problems before solving.

$$^{-}10.6 - {^-}2.4 = {^-}10.6 + 2.4 = {^-}8.2$$

Use arrows on the number line to find the sum or difference. **Check students' number lines.**

1. $^{-}3.2 + 4.5$ ___**1.3**___
2. $^{-}6.3 + 6.1$ ___**$^{-}$0.2**___

3. $2.2 - {^-}3.4$ ___**5.6**___
4. $^{-}2.7 - {^-}4.7$ ___**2**___

5. $^{-}5.4 + {^-}2.1$ ___**$^{-}$7.5**___
6. $^{-}2.3 - 5$ ___**$^{-}$7.3**___

Look at the problems you solved above. Write some rules to help you know how to solve addition problems with rational numbers.

7. The sum of two negative numbers is always ___**negative**___.

8. The sum of two positive numbers is always ___**positive**___.

9. When the signs are different, subtract the absolute values of the two numbers and use the sign of ___**the number with the greater absolute value.**___

TAKE ANOTHER LOOK R23

Name _____

LESSON 5.5

Multiplying and Dividing Rational Numbers

You know about inverse operations. Using inverse operations and your calculator can help you learn the rules for multiplication and division of rational numbers.

SAMPLE KEY SEQUENCES	
Division Problem	**Related Multiplication Problem**
17.94 ÷ ⁻2.3	⁻7.8 × ⁻2.3
17.94 [÷] 2.3 [+/−] [=] ⁻7.8	7.8 [+/−] [×] 2.3 [+/−] [=] 17.94
17.94 ÷ ⁻2.3 = ⁻7.8	⁻7.8 × ⁻2.3 = 17.94

Use your calculator to help you complete the table.

	Division Problem	Related Multiplication Problem
1.	⁻31.25 ÷ 25 = __⁻1.25__	⁻1.25 × 25 = __⁻31.25__
2.	⁻140.7 ÷ __40.2__ = __⁻3.5__	⁻3.5 × 40.2 = ⁻140.7
3.	⁻350 ÷ ⁻70 = __5__	__5__ × __⁻70__ = __⁻350__
4.	__0.5__ ÷ __⁻0.01__ = __⁻50__	⁻50 × ⁻0.01 = 0.5
5.	22.5 ÷ ⁻0.9 = __⁻25__	__⁻25__ × __⁻0.9__ = __22.5__
6.	__⁻4__ ÷ __0.5__ = __⁻8__	⁻8 × 0.5 = __⁻4__

Look at the problems in the table. Write some rules to help you know how to solve multiplication and division problems with rational numbers.

7. When the signs of both factors are negative, the product is __positive__.

8. When the sign of one factor is positive and the other is negative, the product is __negative__.

9. A negative number divided by a positive number is __negative__.

10. A negative number divided by a negative number is __positive__.

R24 TAKE ANOTHER LOOK

Name _____

LESSON 6.1

Numerical and Algebraic Expressions

You use symbols all the time.
$ means *dollar*. ¢ means *cents*. ☺ means *happy*.

You can use different words to represent addition, subtraction, multiplication, and division. For example, *plus* and *added to* are both represented by +.

To write a word expression as an algebraic expression, you need to know how to organize the numbers and variables with the given operation.

Subtract x from 10 is written $10 - x$. ← Make sure that a variable mentioned first in a word expression is placed second in an algebraic expression.

Write the symbol +, −, ×, or ÷ that represents each word or words.

1. times __×__
2. less than __−__
3. increased by __+__
4. quotient __÷__
5. greater than __+__
6. product __×__
7. subtracted from __−__
8. more than __+__
9. divided by __÷__
10. decreased by __−__
11. difference __−__
12. of __×__
13. per __÷__
14. sum __+__
15. factors __×__

For Exercises 16–25, write an algebraic expression that includes a variable, an operation, and a number. **Expressions may vary.**

16. 7 added to a number, *n*
    _____$n + 7$_____

17. the product of 3 and a number, *n*
    _____$3n$_____

18. 9 decreased by a number, *n*
    _____$9 - n$_____

19. a number, *n*, subtracted from 400
    _____$400 - n$_____

20. a number, *c*, divided into 6 equal parts
    _____$c \div 6$_____

21. 20 plus a number, *t*
    _____$20 + t$_____

22. 7.5 greater than a number, *y*
    _____$y + 7.5$_____

23. 30 more than a number, *c*
    _____$c + 30$_____

24. $\frac{2}{3}$ of a number, *c*
    _____$\frac{2}{3}c$_____

25. 5 less than a number, *q*
    _____$q - 5$_____

26. Circle the numerical expression.
27. Underline the algebraic expression.
28. Put a check in front of the variable.

✓ *v*
(1.05 − 0.75)
<u>3*y* − 1</u>

TAKE ANOTHER LOOK R25

LESSON 6.2

Evaluating Expressions

If someone says to you, "Throw me down the stairs my hat," the order of words gives a different instruction from the one intended.

In math, the order of instructions, or operations, is also important.
- The expression 4 + 10 ÷ 2 equals 7 if you do the addition before the division.
- But 4 + 10 ÷ 2 equals 9 if you do the division before the addition.

So, rules exist for the order in which mathematical operations should be performed.

> **Order of Operations**
> 1. Do all operations inside the parentheses.
> 2. Clear the exponents.
> 3. Multiply and divide from left to right.
> 4. Add and subtract from left to right.

Maria says 5 − 4 × 2 = ⁻3. Joe insists that 5 − 4 × 2 = 2. Who is correct? Maria is correct. Joe did not multiply before he subtracted.

When you evaluate an algebraic expression, you replace the variables with the given values. Then you use the order of operations to simplify the expression.

Use the order of operations to evaluate each numerical expression.

1. 13 − 8 − 4   **1**
2. 13 − (8 − 4)   **9**
3. 3 + 4 × 2   **11**
4. 3 × 5 − 1.5   **13.5**
5. $\frac{1}{2}$ × 8 + 6 ÷ 2   **7**
6. ⁻6(4) + 8   **⁻16**
7. 8 + 6 ÷ 2 + 3   **14**
8. (9 + 11) ÷ ⁻4   **⁻5**
9. (2 − 3) + (7 − 5)   **1**

Complete to evaluate each algebraic expression for $a = 3$ and $b = 1.5$.

10. 2a + b − 1
    = 2(3) + 1.5 − 1
    = **6** + 1.5 − 1
    = **7.5** − 1
    = **6.5**

11. a + 1 × 3
    = **3** + 1 × 3
    = **3** + **3**
    = **6**

12. 2b ÷ a + 2²
    = 2(**1.5**) ÷ **3** + **4**
    = **3** ÷ **3** + **4**
    = **1** + **4**
    = **5**

13. 5a + 3b   **19.5**
14. a + 2b ÷ 3   **4**
15. 3 + 2b ÷ 3   **4**

R26 TAKE ANOTHER LOOK

Name _____

LESSON 6.3

Combining Like Terms

You simplify numerical expressions by doing the operations inside parentheses first.
$2(8 - 5) = 2(3) = 6$

How can you simplify $3(x - 7)$?

The Distributive Property is used to multiply a number and an algebraic expression inside parentheses.

Simplify $3(x - 7)$: $= 3(x + {}^-7)$ ← Distribute the "3."
$= 3 \cdot x + 3 \cdot {}^-7$
$= 3x + {}^-21$
$= 3x - 21$ ← Write the expression as a difference without parentheses.

You sometimes use the Distributive Property before simplifying and combining like terms.

$4(x - 3) + 2x = 4 \cdot x + 4 \cdot {}^-3 + 2x = 4x - 12 + 2x = 6x - 12$

Use the Distributive Property to write the expression as a sum or difference without parentheses.

1. $3(y - 9)$ 2. $4(8 + z)$ 3. $5(s + 3)$ 4. $7(q - 6)$

   $\underline{\quad 3y - 27 \quad}$ $\underline{\quad 32 + 4z \quad}$ $\underline{\quad 5s + 15 \quad}$ $\underline{\quad 7q - 42 \quad}$

5. $1.5(n - 7)$ 6. $4(x - 4)$ 7. $8(m - 4.5)$ 8. $4.1(t - 10)$

   $\underline{\quad 1.5n - 10.5 \quad}$ $\underline{\quad 4x - 16 \quad}$ $\underline{\quad 8m - 36 \quad}$ $\underline{\quad 4.1t - 41 \quad}$

Use the Distributive Property to write the expression without parentheses. Then combine like terms to simplify the expression.

9. $7(x + 15) - 5x$ 10. $12g + 3(g + 1)$ 11. $2(t - 3) + 5t$

   $\underline{\quad 7x + 105 - 5x; \quad}$ $\underline{\quad 12g + 3g + 3; \quad}$ $\underline{\quad 2t - 6 + 5t; \quad}$

   $\underline{\quad 2x + 105 \quad}$ $\underline{\quad 15g + 3 \quad}$ $\underline{\quad 7t - 6 \quad}$

12. $2(1.5 - d) + 2.5d$ 13. $3.5h + 4(h - 0.5)$ 14. $5(y - 1) - y$

   $\underline{\quad 3 - 2d + 2.5d; \quad}$ $\underline{\quad 3.5h + 4h - 2; \quad}$ $\underline{\quad 5y - 5 - y; \quad}$

   $\underline{\quad 0.5d + 3 \quad}$ $\underline{\quad 7.5h - 2 \quad}$ $\underline{\quad 4y - 5 \quad}$

15. $5.5(n + 3) - 2.5n$ 16. $6m + 6(3 - m)$ 17. $1.5(x - 6) - 5x$

   $\underline{\quad 5.5n + 16.5 - 2.5n; \quad}$ $\underline{\quad 6m + 18 - 6m; \quad}$ $\underline{\quad 1.5x - 9 - 5x; \quad}$

   $\underline{\quad 3n + 16.5 \quad}$ $\underline{\quad 18 \quad}$ $\underline{\quad {}^-3.5x - 9 \quad}$

TAKE ANOTHER LOOK   R27

Name _____

LESSON 6.4

Sequences and Expressions

Replace the variable in the expression $3n + 2$ with each of the counting numbers, 1, 2, 3, ..., and you discover the pattern 5, 8, 11, 14, ...

$8 - 5 = 3$
$11 - 8 = 3$
$14 - 11 = 3$

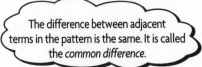
The difference between adjacent terms in the pattern is the same. It is called the common difference.

Because there is a common difference, this pattern is an *arithmetic sequence*.

In the expression $3n + 2$, the $3n$ tells you that the numbers in the sequence will have a common difference of 3. The 2 tells you that each term will be 2 greater than the multiples of 3.

If you replace n with 10, you find the 10th term of the sequence.

$3(10) + 2 = 32$

Suppose you replace the variable in the expression $5n - 1$ with the counting numbers.

1. Write the first five numbers in the sequence. __**4, 9, 14, 19, 24**__

2. What is the common difference? __**5**__

3. How do the numbers in the sequence differ from the multiples of 5?
   __**Terms are 1 less than multiples of 5.**__

4. What is the 10th term in the sequence? __**49**__

You can write an expression to describe the sequence 8, 10, 12, 14, 16, ...

- The common difference is 2. So, the first term in the expression is $2n$.
- The first term 8 is 6 more than the first multiple of 2. So, the second part of the expression is +6.
- The expression $2n + 6$ describes the sequence.

5. Write the first five terms in an arithmetic sequence where the first term is 5 and the common difference is 4. __**5, 9, 13, 17, 21**__

6. Write an expression to describe the sequence. __**4n + 1**__

7. Write the first five terms in an arithmetic sequence where the first term is 5 and the common difference is 5. __**5, 10, 15, 20, 25**__

8. Write an expression to describe the sequence. __**5n**__

R28 TAKE ANOTHER LOOK

Connecting Equations and Words

Lesson 7.1

Complete the table.

Look for these clue words to help you write sentences as equations.

Clue Words for Addition	Clue Words for Subtraction	Clue Words for Multiplication	Clue Words for Division
• add • and • more than • increased by • plus • sum	• subtract • decreased by • less than • fewer than • minus • difference • less	• times • multiplied by • twice • double • product • of	• divided • split evenly • quotient • per

	Sentence	Clue Word(s)	Operation	Equation
	5 less than x is 15.	less than	subtraction	$x - 5 = 15$
1.	r divided by 8 is 9.	divided by	division	$\frac{r}{8} = 9$
2.	9 increased by m is 15.	increased by	addition	$9 + m = 15$
3.	p minus 3 is 12.	minus	subtraction	$p - 3 = 12$
4.	Six times t is 42.	times	multiplication	$6t = 42$
5.	g split evenly by 2 is 11.	split evenly	division	$\frac{g}{2} = 11$
6.	Eight more than s is 24.	more than	addition	$s + 8 = 24$
7.	4 fewer than n is 20.	fewer than	subtraction	$n - 4 = 20$
8.	c divided by 2 is 32.	divided by	division	$\frac{c}{2} = 32$
9.	w and 14 are 26.	and	addition	$w + 14 = 26$
10.	Twice j is 18.	twice	multiplication	$2j = 18$
11.	The quotient of x and 2 is 5.	quotient	division	$\frac{x}{2} = 5$
12.	$\frac{1}{3}$ of n is 4.	of	multiplication	$\frac{1}{3}n = 4$
13.	5 subtracted from x is $^-2$.	subtracted from	subtraction	$x - 5 = {}^-2$
14.	n less than 3 is 2.	less than	subtraction	$3 - n = 2$
15.	3 subtracted from b is 5.	subtracted from	subtraction	$b - 3 = 5$

TAKE ANOTHER LOOK

Name _____

LESSON 7.2

Solving Addition and Subtraction Equations

How do you find what value of the variable makes a statement true?

To solve an equation, you need to arrange its parts so that the variable is alone on one side of the equal sign.

To do that, use inverse operations.

Inverse operations "undo" each other.
- The inverse of addition is subtraction.
- The inverse of subtraction is addition.

Solve.	$b + 14 = 27$	How do you undo "+ 14"?
	$b + 14 - \mathbf{14} = 27 - \mathbf{14}$	Subtract 14 from each side of the equation.
	$b = 13$	Since the variable is alone on one side, you have solved the equation.

Solve.	$p - 35 = 19$	How do you undo "− 35"?
	$p - 35 + \mathbf{35} = 19 + \mathbf{35}$	Add 35 to each side of the equation.
	$p = 54$	Since the variable is alone on one side, you have solved the equation.

Remember, when you add or subtract a number from one side of an equation, do the same to the other side to keep the sides equal.

Complete, showing the steps for solving the equation.

1. $m - 14 = 59$

   $m - 14 + \underline{\ 14\ } = 59 + \underline{\ 14\ }$

   $m = \underline{\ 73\ }$

2. $r + 17 = 41$

   $r + 17 - \underline{\ 17\ } = 41 - \underline{\ 17\ }$

   $r = \underline{\ 24\ }$

3. $t + 3 = 7$

   $t + 3 - \underline{\ 3\ } = 7 - \underline{\ 3\ }$

   $t = \underline{\ 4\ }$

4. $n - 1 = 4$

   $n - 1 + \underline{\ 1\ } = 4 + \underline{\ 1\ }$

   $n = \underline{\ 5\ }$

5. $s - 6 = 10$

   $s - 6 + \underline{\ 6\ } = 10 + \underline{\ 6\ }$

   $s = \underline{\ 16\ }$

6. $x + 2 = 4$

   $x + 2 - \underline{\ 2\ } = 4 - \underline{\ 2\ }$

   $x = \underline{\ 2\ }$

Solve.

7. $d + 5 = 9$

   $d = \underline{\ 4\ }$

8. $j - 3 = 12$

   $j = \underline{\ 15\ }$

R30 **TAKE ANOTHER LOOK**

LESSON 7.3

Multiplication and Division Equations

Find the value of the variable. Use inverse operations to arrange the equation so that the variable is alone on one side of the equation.

Division and multiplication are inverse operations. Division undoes multiplication, and multiplication undoes division.

Solve.	$14b = 84$ $\quad \dfrac{14b}{14} = \dfrac{84}{14}$ $\quad b = 6$	Use division to undo multiplication. Divide each side by 14. When the variable is alone on one side, the equation is solved.
Solve.	$\dfrac{c}{5} = 21$ $\quad 5 \times \dfrac{c}{5} = 21 \times 5$ $\quad c = 105$	Use multiplication to undo division. Multiply each side by 5. When the variable is alone on one side, the equation is solved.

Remember, when you multiply or divide one side of an equation by a number, do the same to the other side to keep the sides equal.

Complete, showing the steps for solving the equation.

1. $9c = 144$

   $\dfrac{9c}{\boxed{9}} = \dfrac{144}{\boxed{9}}$

   $c = \underline{\ 16\ }$

2. $13t = 104$

   $\dfrac{13t}{\boxed{13}} = \dfrac{104}{\boxed{13}}$

   $t = \underline{\ 8\ }$

3. $2m = 14$

   $\dfrac{2m}{\boxed{2}} = \dfrac{14}{\boxed{2}}$

   $m = \underline{\ 7\ }$

4. $\dfrac{m}{9} = 4$

   $\dfrac{\boxed{9 \times}\, m}{9} = 4\ \underline{\times 9}$

   $m = \underline{\ 36\ }$

5. $\dfrac{s}{3} = 2$

   $\dfrac{\boxed{3 \times}\, s}{3} = 2\ \underline{\times 3}$

   $s = \underline{\ 6\ }$

6. $\dfrac{p}{1} = 6$

   $\dfrac{\boxed{1 \times}\, p}{1} = 6\ \underline{\times 1}$

   $p = \underline{\ 6\ }$

7. $2b = 28$

   $\dfrac{2b}{\boxed{2}} = \dfrac{28}{\boxed{2}}$

   $b = \underline{\ 14\ }$

8. $4r = 44$

   $\dfrac{4r}{\boxed{4}} = \dfrac{44}{\boxed{4}}$

   $r = \underline{\ 11\ }$

9. $\dfrac{k}{2} = 4$

   $\dfrac{\boxed{2 \times}\, k}{2} = 4\ \underline{\times 2}$

   $k = \underline{\ 8\ }$

TAKE ANOTHER LOOK

Name _____

LESSON 7.4

Problem-Solving Strategy

Working Backward to Solve Problems

Alice received her weekly paycheck. She put $\frac{1}{2}$ in a savings account. Then she bought some tapes for $15. She has $8 left. How much money did Alice get in her paycheck?

Step 1 Find how much money Alice had *after* she put $\frac{1}{2}$ of her paycheck in the bank.
- She has $8 after spending $15 on tapes.
- The sum $8 + $15 is how much money she had after depositing $\frac{1}{2}$ of her paycheck.
- $8 + $15 = $23

Step 2 Since $\frac{1}{2}$ of Alice's paycheck is $23, the total amount of the paycheck is 2 times $23.
- $23 × 2 = $46

Alice's paycheck was for $46.

Work backward to solve. Complete Problems 1 and 2 to show the steps.

1. Vince received his weekly paycheck. He deposited $\frac{1}{2}$ in the bank. Then Vince spent $12 at the bookstore and bought a pizza for $8. If Vince has $14 left, what was the amount of his paycheck?

   Step 1 Find how much money Vince had after putting $\frac{1}{2}$ of the check in the bank.
   - Add. __12__ + __8__ + __14__ = __34__

   Step 2 The sum is $\frac{1}{2}$ of Vince's paycheck. Multiply __34__ by __2__ to find the total amount of the paycheck.

   Vince's paycheck was for __$68__.

2. Hillary receives a monthly allowance. This month she put $\frac{1}{4}$ of her allowance in the bank. Then she spent $23 on clothes and $2 for lunch. She has $5 left. How much allowance does Hillary receive each month?

   Step 1 Find how much money Hillary had after putting $\frac{1}{4}$ of her allowance in the bank.
   - Add. __23__ + __2__ + __5__ = __30__

   Step 2 The sum is $\frac{3}{4}$ of her total allowance.
   - To find $\frac{1}{4}$ of Hillary's total allowance, divide. __30__ ÷ 3 = __10__
   - To find $\frac{4}{4}$ of Hillary's total allowance, multiply. __10__ × 4 = __40__

   Hillary receives an allowance of __$40__ each month.

R32 **TAKE ANOTHER LOOK**

Name _____

LESSON 7.5

Proportions

A cyclist travels 28 mi in 2 hr. At the same rate, how far would she travel in 3.5 hr?

Step 1 THINK: What is being compared? The cyclist's *distance* is compared with a period of *time*.

So, your ratios should both be $\frac{distance}{time}$.

Step 2 Write the ratios in your proportion.

$\text{distance} \rightarrow \frac{28 \text{ mi}}{2 \text{ hr}} = \frac{d}{3.5 \text{ hr}}$

Step 3 Write the cross products.

$\frac{28}{2} = \frac{d}{3.5}$

$28 \times 3.5 = 2 \times d$

Step 4 Multiply.

$98 = 2d$

Step 5 Divide to solve the equation.

$\frac{98}{2} = \frac{2d}{2}$

The cyclist would travel 49 mi in 3.5 hr.

$49 = d$, or $d = 49$

---

1. A proportion is an equation that states that two ___**ratios**___ are equivalent.

Complete. Variables may vary.

2. The Pizza Palace sold 27 pizzas in 30 min. At this rate, how many pizzas would be sold in 240 min?

   What is being compared? ___**pizzas to minutes**___

   So, your ratios should both be ___**pizzas/minutes**___.

   What proportion do you use to solve the problem? ___$\frac{27}{30} = \frac{p}{240}$___

   Write the cross products. ___$27 \times 240 = 30p$; $6{,}480 = 30p$___

   What is the solution of the equation? ___$p = 216$___

   How many pizzas would be sold in 240 minutes? ___**216 pizzas**___

Solve.

3. Mary works at Freezie's. On Friday she uses 8 qt of ice cream to make 100 milk shakes. She expects to make 550 milk shakes on Saturday. How many quarts of ice cream will she need to make 550 milk shakes?

   ___**44 qt**___

4. The ratio of Rafael's weight on earth to his weight on the moon is 6 to 1.

   Rafael weighs 90 lb. How much would he weigh on the moon? ___**15 lb**___

TAKE ANOTHER LOOK R33

Name _____

Problem-Solving Strategy

LESSON 8.1

Write an Equation to Solve Two-Step Problems

Mia and Tim are on the school soccer team. Mia has scored 5 more than twice the number of goals that Tim has scored. She has scored a total of 23 goals. How many goals has Tim scored?

You can write a two-step equation to solve this problem.

- Use numbers and symbols to represent words.
- Let n represent the number of goals scored by Tim.

5 more than	twice the number of goals Tim scored	equals	the number of goals Mia scored
5 +	2n	=	23

- Write a two-step equation. Because addition is commutative, you can write the equation as $2n + 5 = 23$.
- Solve the equation.

Remember, your goal is to get the variable n alone on one side of the equation.

$$2n + 5 = 23$$
$$2n + 5 - 5 = 23 - 5$$
$$\frac{2n}{2} = \frac{18}{2}$$
$$n = 9$$

- Replace n in the original equation to check.

Complete the steps to write a two-step equation to solve each problem. Then find the solution. **Check students' equations. Variables may vary.**

1. Ted saved $5.30 more than twice the amount Mel saved. Ted's savings total $27.80. How much has Mel saved?

   - Represent the amount Mel has saved with __x__.
   - Represent *twice the amount Mel saved* with __2x__.
   - Represent *$5.30 more than twice the amount* with $2x + 5.30$.
   - Represent *equals* with __=__.
   - Write a two-step equation. Then solve it to solve the problem.

   $\underline{2x + 5.3 = 27.8; \ x = 11.25; \ \$11.25}$

2. The number of CDs Ron has is 8 fewer than three times the number of CDs Chris has. If Ron has 28 CDs, how many does Chris have?

   - Represent the number of CDs Chris has with __c__.
   - Represent *three times the number of CDs Chris has* with __3c__.
   - Represent *8 fewer than three times the number* with $3c - 8$.
   - Represent *is* with __=__.
   - Write a two-step equation. Then solve it to solve the problem.

   $\underline{3c - 8 = 28; \ c = 12; \ 12 \text{ CDs}}$

R34 TAKE ANOTHER LOOK

Lesson 8.2

Simplifying and Solving

Sometimes you need to use the Distributive Property to simplify an equation before solving. Parentheses are a clue that you may need to use the Distributive Property.

Solve and check. $5(2t - 6) + 5t = 75$

Multiply each term inside the parentheses by the number outside the parentheses. (Distributive Property)	$5(2t - 6) + 5t = 75$ $(5 \times 2t) - (5 \times 6) + 5t = 75$
Simplify by combining like terms.	$10t - 30 + 5t = 75$ $15t - 30 = 75$
Solve the simplified equation.	$15t - 30 + 30 = 75 + 30$ $\dfrac{15t}{15} = \dfrac{105}{15}$ $t = 7$
Check your solution in the simplified equation.	$15t - 30 = 75$ $15(7) - 30 \stackrel{?}{=} 75$ $105 - 30 \stackrel{?}{=} 75$ $75 = 75$ ✓

Complete. Be sure to check your solution.

1. $2(3r + 8) + 11r = 118$

   $(\underline{2} \times \underline{3r}) + (\underline{2} \times \underline{8}) + 11r = 118$

   $\underline{6r} + \underline{16} + 11r = 118$

   $\underline{17r} + \underline{16} = 118$

   $\underline{17r} + \underline{16} - \underline{16} = 118 - \underline{16}$

   $17r = 102$

   $\dfrac{\boxed{17r}}{\boxed{17}} = \dfrac{\boxed{102}}{\boxed{17}}$

   $\underline{r} = \underline{6}$

Solve.

2. $2.4(3b + 1.5) + 1.8b = 44.1$

   $\dfrac{\boxed{9b}}{\boxed{9}} = \dfrac{\boxed{40.5}}{\boxed{9}}$

   $\underline{b} = \underline{4.5}$

3. $1.2(5 + 3p) + 7.4p - 2.5 = 15.6$

   $\dfrac{\boxed{11p}}{\boxed{11}} = \dfrac{\boxed{12.1}}{\boxed{11}}$

   $\underline{p} = \underline{1.1}$

TAKE ANOTHER LOOK

Name _____

LESSON 8.3

Comparing Equations and Inequalities

An *inequality* is a kind of number sentence. It describes a relationship between two *unequal* quantities. Five different symbols can be used in an inequality. Each symbol shows a different relationship.

SYMBOL	MEANING	EXAMPLE
$<$	is less than	$t < 6$
$>$	is greater than	$x > 10$
$\leq$	is less than or equal to	$b \leq 5$
$\geq$	is greater than or equal to	$m \geq 1$
$\neq$	is not equal to	$y \neq 4$

Replace the variable with the given value or values. Then tell which inequality symbol should complete the sentence: $>$, $<$, $\geq$, or $\leq$.

1. $t + 5 \bigcirc 6$ $t = 2$ $\underline{\;>\;}$
2. $b - 1 \bigcirc 8$ $b = 5$ $\underline{\;<\;}$
3. $c + 1 \bigcirc 5$ $c = 3$ $\underline{\;<\;}$
4. $f - 5 \bigcirc 8$ $f = 10$ $\underline{\;<\;}$
5. $y - 4 \bigcirc 11$ $y = 15$ or 18 $\underline{\;\geq\;}$
6. $g + 7 \bigcirc 9$ $g = 0$ or 2 $\underline{\;\leq\;}$
7. $b + 3 \bigcirc 5$ $b = 1$ or 2 $\underline{\;\leq\;}$
8. $2c - 1 \bigcirc 4$ $c = 2$ $\underline{\;<\;}$
9. $4m + 2 \bigcirc 5$ $m = 3$ $\underline{\;>\;}$
10. $9x - 6 \bigcirc 3$ $x = 4$ $\underline{\;>\;}$
11. $5n + 1.6 \bigcirc 6.6$ $n = 1$ or 3 $\underline{\;\geq\;}$
12. $3a \bigcirc 15$ $a = 1.1$ $\underline{\;<\;}$
13. $4m \bigcirc 4^2$ $m = 5$ $\underline{\;>\;}$
14. $x \bigcirc \frac{1}{2} - \frac{1}{2}$ $x = {}^-1$ $\underline{\;<\;}$

R36   TAKE ANOTHER LOOK

Name _____

LESSON 8.4

Solving Inequalities

The prefix *in-* means *not*. An *in*equality is a number sentence that describes two quantities that are *not* equal.

You solve an inequality the same way you solve an equation. The key is always to perform the same operation on both sides of the inequality.

Solve. $b - 4 > 1$ • You read this as "the difference between b and 4 is greater than 1."

$b - 4 + 4 > 1 + 4$ • Add 4 to both sides of the inequality.

$b > 5$ • The solution is any number greater than 5.

Graph the solution.
- Locate 5 on the number line.
- Put an *open* circle on 5, because 5 is *not* a solution.
- Draw a ray to the *right* of 5, because all solutions are *greater than* 5.

Complete the steps to solve each inequality. Graph the solution. **Check students' graphs.**

1. $m + 2 < 6$
    $m + 2 - \underline{\ 2\ } < 6 - \underline{\ 2\ }$
    $m < \underline{\ 4\ }$

2. $n - 8 > 3$
    $n - 8 + \underline{\ 8\ } > 3 + \underline{\ 8\ }$
    $n > \underline{\ 11\ }$

3. $2t - 1 \geq 7$
    $2t - 1 + \underline{\ 1\ } \geq 7 + \underline{\ 1\ }$
    $2t \geq \underline{\ 8\ }$
    $\dfrac{2t}{\boxed{2}} \geq \dfrac{\boxed{8}}{\boxed{2}}$
    $t \geq \underline{\ 4\ }$

4. $1.4r - 2.1 \leq 4.9$
    $1.4r - 2.1 + \underline{\ 2.1\ } \leq 4.9 + \underline{\ 2.1\ }$
    $1.4r \leq \underline{\ 7\ }$
    $\dfrac{1.4r}{\boxed{1.4}} \leq \dfrac{\boxed{7}}{\boxed{1.4}}$
    $r \leq \underline{\ 5\ }$

5. $\dfrac{x}{3} + 2 > {}^-4$
    $\dfrac{x}{3} + 2 - \underline{\ 2\ } > {}^-4 - \underline{\ 2\ }$
    $\dfrac{x}{3} > \underline{\ {}^-6\ }$
    $\dfrac{x}{3} \times \underline{\ 3\ } > \underline{\ {}^-6\ } \times \underline{\ 3\ }$
    $x > \underline{\ {}^-18\ }$

6. $\dfrac{x}{4} + 1 \geq 1$
    $\dfrac{x}{4} + 1 - \underline{\ 1\ } \geq 1 - \underline{\ 1\ }$
    $\dfrac{x}{4} \geq \underline{\ 0\ }$
    $\dfrac{x}{4} \times \underline{\ 4\ } \geq \underline{\ 0\ } \times \underline{\ 4\ }$
    $x \geq \underline{\ 0\ }$

TAKE ANOTHER LOOK R37

Name _____

LESSON 9.1

Graphing Ordered Pairs

An *ordered pair* shows the location of a point on a coordinate plane.

The first number, the *x*-coordinate, shows the location of a point along the *x*-axis.

- If the *x*-coordinate is positive, the point is to the right of the origin.
- If the *x*-coordinate is negative, the point is to the left of the origin.

The second number of an ordered pair, the *y*-coordinate, shows the location of a point along the *y*-axis.

- If the *y*-coordinate is positive, the point is above the origin.
- If the *y*-coordinate is negative, the point is below the origin.

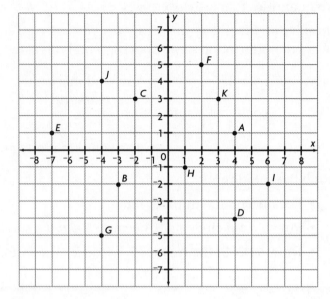

What are the *x*-coordinates and *y*-coordinates of points *A* and *B*?

The *x*-coordinate of point *A* is 4. The *y*-coordinate of point *A* is 1.

The *x*-coordinate of point *B* is ⁻3. The *y*-coordinate of point *B* is ⁻2.

Complete, using the coordinate plane above.

1. point *C*

    x-coordinate ___⁻2___

    y-coordinate ___3___

    ordered pair ___(⁻2,3)___

2. point *D*

    x-coordinate ___4___

    y-coordinate ___⁻4___

    ordered pair ___(4,⁻4)___

3. point *E*

    x-coordinate ___⁻7___

    y-coordinate ___1___

    ordered pair ___(⁻7,1)___

4. point *F*

    x-coordinate ___2___

    y-coordinate ___5___

    ordered pair ___(2,5)___

5. point *G*

    x-coordinate ___⁻4___

    y-coordinate ___⁻5___

    ordered pair ___(⁻4,⁻5)___

6. point *H*

    x-coordinate ___1___

    y-coordinate ___⁻1___

    ordered pair ___(1,⁻1)___

7. point *I*

    x-coordinate ___6___

    y-coordinate ___⁻2___

    ordered pair ___(6,⁻2)___

8. point *J*

    x-coordinate ___⁻4___

    y-coordinate ___4___

    ordered pair ___(⁻4,4)___

9. point *K*

    x-coordinate ___3___

    y-coordinate ___3___

    ordered pair ___(3,3)___

Name _____

LESSON 9.2

Relations

A graph and a table of ordered pairs are two ways to show a relation. A relation matches one set of elements, called the *domain*, with a second set of elements, called the *range*.

This graph shows a relation. You can make a table of ordered pairs to show this relation.

Begin by looking at point *A*. The *x*-coordinate is 1. The *y*-coordinate is 2. The ordered pair that names point *A* is (1,2).

1. Complete the table for the other points.

Point	x-coordinate	y-coordinate	Ordered Pair
A	1	2	(1,2)
B	2	4	(2,4)
C	3	6	(3,6)
D	4	8	(4,8)

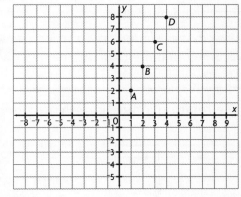

The *x*-coordinates make up the domain of a relation. The *y*-coordinates make up the range.

2. What are the domain and range of the above relation?

   domain = {1, 2, 3, 4}; range = {2, 4, 6, 8}

You can also express a relation as an equation. In the above relation, the *y*-coordinate equals the *x*-coordinate multiplied by 2. So, the equation that represents this relation is $y = 2x$.

For Exercises 3–5, use the graph at the right.

3. Complete the table for the relation shown.

Point	x-coordinate	y-coordinate	Ordered Pair
E	0	⁻2	(0,⁻2)
F	1	⁻1	(1,⁻1)
G	2	0	(2,0)
H	3	1	(3,1)

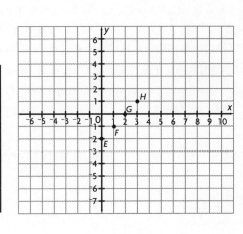

4. What are the domain and range of the relation?

   domain = {0, 1, 2, 3}; range = {⁻2, ⁻1, 0, 1}

5. Write an equation to represent the relation. _____ $y = x - 2$ _____

TAKE ANOTHER LOOK R39

Name _____

LESSON 9.3

Functions

A function is a special relation in which each element of the domain maps to *one* and *only one* member of the range.

Is this relation a function?

To find if a relation is a function, look at the domain. Check to see that each member of the domain is mapped to *only one* member of the range.

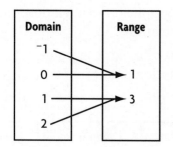

This relation is a function. Each member of the domain maps to only one member of the range.

Notice that a relation is a function even if a member of the range is mapped to more than one member of the domain.

Is the relation a function? Write *yes* or *no*. If you write *no*, explain.

1. 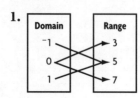 No; zero in the domain maps to two range numbers.

2. 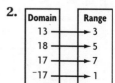 yes

3. 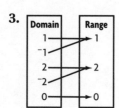 yes

4. Domain/Range diagram No; 1, 2, and 3 in the domain map to more than one number of the range.

You can use the vertical line test to see if a relation is a function. Check if a vertical line drawn on the graph crosses two or more points. If so, an *x*-coordinate maps to two or more members of the range, so the relation is *not* a function.

Is this relation a function? Can you draw a vertical line on the graph that will cross two or more points?

Yes; 2 in the domain maps to four members of the range. This is not a function.

Use the vertical line test to determine if the relation is a function.

5. 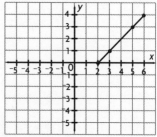 yes

6. 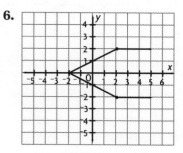 no

R40 **TAKE ANOTHER LOOK**

Name _____

LESSON 9.4

Linear Equations

You can find solutions to an equation with two variables.
- Choose a value for one variable.
- Substitute the value in the equation.
- Solve the equation to find the value of the other variable.

Find solutions of the equation $r = 3b$. Then graph the equation.

Step 1
Make a table of values. Choose a value for b. Find the value of $3b$. This is the value for r. Repeat with different values for b.

Step 2
Write the solutions as ordered pairs.

Step 3
Graph the ordered pairs. Draw a line through the ordered pairs.

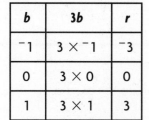
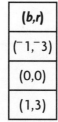
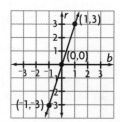

1. Complete to find solutions of the equation $t = 4 - m$. Then graph the equation. **Check students' graphs.**

   Step 1

m	4 − m	t
1	4 − 1	3
2	4 − 2	2
3	4 − 3	1

   Step 2

(m,t)
(1,3)
(2,2)
(3,1)

   Step 3
   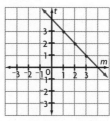

2. Complete to find solutions of the equation $y = 3x - 1$. Then graph the equation. **Check students' graphs. Ordered pairs may vary.**

   Step 1

x	3x − 1	y
1	3 × 1 − 1	2
2	3 × 2 − 1	5
3	3 × 3 − 1	8

   Step 2

   Step 3
   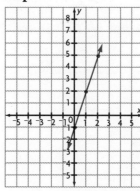

TAKE ANOTHER LOOK   R41

Name _____

LESSON 10.1

Congruent Line Segments and Angles

You can use what you know about congruent line segments and angles to solve problems with isosceles triangles.

Triangle ABC is isosceles. The measure of ∠CBA is 70°. Find the measures of ∠CAB and ∠BCA.

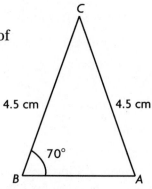

Step 1 Because △ABC is isosceles, the base angles are congruent. ∠CBA measures 70°, so ∠CAB measures 70°.

Step 2 The sum of the measures of the angles of a triangle is 180°.
m∠CBA + m∠CAB + m∠BCA = 180°
Substitute: 70° + 70° + m∠BCA = 180°
140° + m∠BCA = 180°
140° − 140° + m∠BCA = 180° − 140°
∠BCA measures 40°.

The measures of ∠CAB and ∠BCA are 70° and 40°.

For Exercises 1–2, complete.

1. Triangle MPN is isosceles. The measure of ∠MNP is 35°. Find the measures of ∠NMP and ∠MPN.

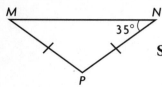

   Step 1 Because △MPN is isosceles, the base angles are ___congruent___.

   ∠MNP is congruent to ___∠NMP___. ∠NMP measures ___35°___.

   Step 2 The sum of the measures of the angles of a triangle

   is ___180°___. The sum of the base angles is ___70°___,

   so ∠MPN measures ___110°___.

2. Triangle WXY is isosceles. The measure of ∠XWY is 55°. Find the measures of ∠XYW and ∠WXY.

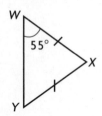

   Step 1 ∠XWY is congruent to ___∠XYW___, so ∠XYW measures ___55°___.

   Step 2 The sum of the base angles is ___110°___, so ∠WXY

   measures ___70°___.

Write the correct letter from Column 2.

3. adjacent angles   ___c___

4. common vertex   ___d___

                         a. angles that contain the same number of degrees

                         b. opposite angles when two lines intersect

5. congruent angles   ___a___

                         c. angles having a common vertex and a common side

                         d. vertical angles and adjacent angles always have this

6. vertical angles   ___b___

R42 TAKE ANOTHER LOOK

Symmetry

A figure has rotational symmetry if it can be rotated, or turned, less than 360° about a central point and match the original figure.

A regular polygon has congruent sides and angles. A regular polygon has rotational symmetry.

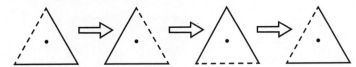

An equilateral triangle has rotational symmetry. Since it has 3 sides, an equilateral triangle has $\frac{1}{3}$-turn or 360° ÷ 3 = 120° symmetry.

Look at these regular polygons. Complete to describe the rotational symmetry of each polygon.

1. A(n) __square__ has rotational symmetry. Since it has __4__ sides, a square has $\frac{1}{4}$-turn or __90°__ symmetry.

2.

   A regular __pentagon__ has rotational symmetry. Since it has __5__ sides, a pentagon has $\frac{1}{5}$-turn or __72°__ symmetry.

3. A regular __octagon__ has rotational symmetry. Since it has __8__ sides, an octagon has $\frac{1}{8}$-turn or __45°__ symmetry.

Match each word with its description.

4. line symmetry __b__   a. when a figure matches the original figure after being rotated less than 360° about a central point

5. point of rotation __c__   b. when two halves of a figure match when folded on a line

6. rotational symmetry __a__   c. central point of a figure that has rotational symmetry

Transformations

LESSON 10.3

Look at a newspaper reflected in a mirror, the blade of a windmill blowing in the wind, or your math book moved to your friend's desk. These are examples of transformations.

In a *reflection,* a figure is flipped over a line, like the reflected newspaper. Both the position and location of the figure change.

In a *rotation,* a figure is turned clockwise or counterclockwise about a point, like the blade of the windmill. Both the position and location of the figure change.

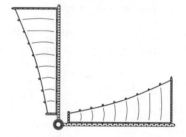

In a *translation,* a figure is slid to a new location, like the math book. The position of the figure remains the same.

Describe the transformation as a *reflection,* a *rotation,* or a *translation.*

1.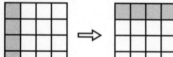

    rotation

2.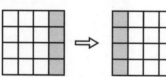

    reflection

3.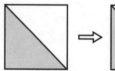

    translation

4.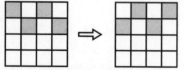

    reflection

5.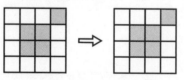

    translation

6.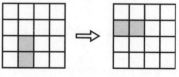

    rotation

R44 TAKE ANOTHER LOOK

Name _____

LESSON 10.4

Transformations on the Coordinate Plane

Look at the triangle. The coordinates of the vertices are A (0,3), B (6,0), and C (0,0).

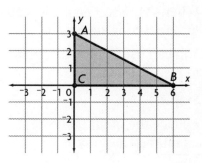

Translate the triangle 4 units left and 2 units up.

Step 1 Slide the triangle 4 units to the left along the x-axis.
- Moving horizontally to the left is moving in a "negative" direction.
- Subtract 4 units from the x-coordinate of each vertex.

  $(0,3) \rightarrow (^-4,3)$

  $(6,0) \rightarrow (2,0)$

  $(0,0) \rightarrow (^-4,0)$

Step 2 Slide the triangle 2 units up.
- Moving up is moving in a "positive" direction.
- Add 2 units to the y-coordinate of each vertex.

  $(^-4,3) \rightarrow (^-4,5)$

  $(2,0) \rightarrow (2,2)$

  $(^-4,0) \rightarrow (^-4,2)$

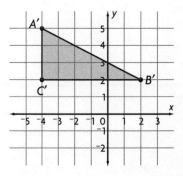

Complete the statements to find the coordinates of each new triangle.

1. The coordinates of the triangle are ___**A (1,⁻1), B (5, 1), and C (1, 2)**___ .
   Translate the triangle 1 unit right and 2 units down.

   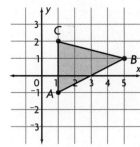

   - Add __1__ to each x-coordinate.
   - Subtract __2__ from each y-coordinate.
   - The coordinates of the new triangle are

     ___**A′ (2,⁻3), B′ (6,⁻1), and C′ (2,0)**___ .

2. The coordinates of the triangle are ___**A (⁻1, 3), B (⁻3,⁻1), and C (0,⁻1)**___ .
   Translate the triangle 3 units left and 2 units up.

   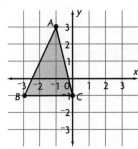

   - Subtract __3__ from each x-coordinate.
   - Add __2__ to each y-coordinate.
   - The coordinates of the new triangle are

     ___**A′ (⁻4,5), B′ (⁻6,1), and C′ (⁻3,1)**___ .

TAKE ANOTHER LOOK R45

Name _____

LESSON 11.1

Constructing Congruent Angles and Line Segments

When you *bisect* an angle, you divide it into two congruent parts.

You can use a compass and a straightedge to bisect ∠JKL.

Step 1
- Place the point of the compass at point K.
- Draw an arc through $\vec{KJ}$ and $\vec{KL}$.
- Label the points of intersection of the arc and the two segments A and B.

Step 2
- Place the point of the compass at A and draw an arc.
- Place the point of the compass at B and draw an arc.
- Label the point of intersection of the two arcs C.

Step 3
- Draw $\vec{KC}$. $\vec{KC}$ is the bisector of ∠JKL.
- So, ∠JKC ≅ ∠CKL.

---

1. Complete to explain how you construct a bisector of ∠ABC. Follow the steps to construct the bisector.

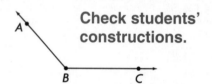

Check students' constructions.

Step 1
- Place the point of the compass at point __B__.
- Draw a(n) __arc__ through $\vec{BA}$ and __$\vec{BC}$__.
- Label the points of __intersection__ D and E.

Step 2
- Place the point of the compass at point __D__ and draw an arc.
- Place the point of the compass at point __E__ and draw an arc.
- Label the point of intersection __F__.

Step 3
- Draw $\vec{BF}$.
- $\vec{BF}$ is the __bisector__ of ∠ABC.
- ∠ABF ≅ __∠FBC__.

2. Construct the bisector of ∠MNO. Explain your steps.

__Check students' constructions and explanations.__

_____

_____

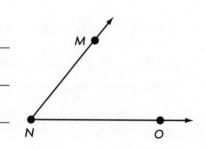

R46 **TAKE ANOTHER LOOK**

Name _____

LESSON 11.2

Constructing Parallel and Perpendicular Lines

Lines, rays, or segments that intersect to form right angles are *perpendicular*.
You can use a compass and a straightedge to construct a perpendicular line.

Step 1
- Place the compass point on any point P on $\overleftrightarrow{MN}$.
- Using the same compass opening, draw arcs on each side of P that intersect $\overleftrightarrow{MN}$.
- Label the points of intersection R and S.

Step 2
- Open the compass wider than $\overline{RP}$.
- With the compass point on R, draw an arc above $\overleftrightarrow{MN}$.
- With the compass point on S, draw an arc above $\overleftrightarrow{MN}$.
- Label the point of intersection T.

Step 3
- Draw $\overrightarrow{PT}$.
- $\overleftrightarrow{MN} \perp \overrightarrow{PT}$.

1. Complete to explain how you construct a line perpendicular to $\overleftrightarrow{AB}$. Follow the steps to construct the line.

Step 1
- Place the point of the

  ___compass___ on any

  point P on ___$\overleftrightarrow{AB}$___.

- Draw two ___arcs___ the same distance from P

  that ___intersect___

  $\overleftrightarrow{AB}$.

- Label the points of

  ___intersection___

  C and D.

Step 2
- Open the compass

  wider than ___$\overline{CP}$___.

- From C and D, draw two

  ___arcs___ that

  ___intersect___ at

  point E.

Step 3
- Draw ___$\overrightarrow{PE}$___.

- Line ___$\overrightarrow{PE}$___ is

  ___perpendicular___

  to $\overleftrightarrow{AB}$ at point P.

- ___$\overleftrightarrow{AB}$___ $\perp$ ___$\overrightarrow{PE}$___.

2. Construct a line perpendicular to $\overleftrightarrow{JK}$. Explain your steps.

___Check students' constructions and explanations.___

TAKE ANOTHER LOOK R47

Name _____

LESSON 11.3

Classifying and Comparing Triangles

Two triangles are congruent when they have the same size and shape.

Fold your paper along a line halfway between the triangles. The vertices and sides of △STR and △XWY match because △STR ≅ △XWY.

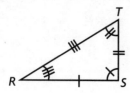 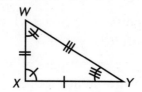

Angles are congruent:
∠S ≅ ∠X
∠T ≅ ∠W
∠R ≅ ∠Y

Sides are congruent:
$\overline{ST} \cong \overline{XW}$
$\overline{TR} \cong \overline{WY}$
$\overline{RS} \cong \overline{YX}$

Complete to match the vertices and sides of the congruent triangles. Then write a congruence statement for the triangles.

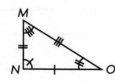 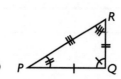   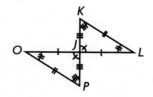

1. a. ∠N ≅ __∠Q__   b. $\overline{NO} \cong$ __QP__   2. a. ∠J ≅ __∠J__   b. $\overline{JO} \cong$ __JL__

   ∠M ≅ __∠R__ $\overline{MN} \cong$ __RQ__   ∠O ≅ __∠L__ $\overline{OP} \cong$ __LK__

   ∠O ≅ __∠P__ $\overline{MO} \cong$ __RP__   ∠P ≅ __∠K__ $\overline{PJ} \cong$ __KJ__

   c. △NMO ≅ __△QRP__   c. △JOP ≅ __△JLK__

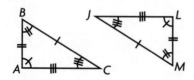   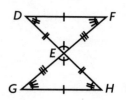

3. a. ∠L ≅ __∠A__   b. $\overline{JL} \cong$ __CA__   4. a. ∠D ≅ __∠H__   b. $\overline{DE} \cong$ __HE__

   ∠J ≅ __∠C__ $\overline{JM} \cong$ __CB__   ∠F ≅ __∠G__ $\overline{EF} \cong$ __EG__

   ∠M ≅ __∠B__ $\overline{LM} \cong$ __AB__   ∠E ≅ __∠E__ $\overline{FD} \cong$ __GH__

   c. △JLM ≅ __△CAB__   c. △DEF ≅ __△HEG__

Name _____

LESSON 11.4

Constructing Congruent Triangles

You construct congruent angles and line segments when you construct congruent triangles.

You can use three rules, Side-Angle-Side, Angle-Side-Angle, or Side-Side-Side, to construct a triangle congruent to △HLM.

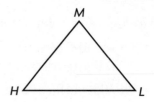

Use SAS	Use ASA	Use SSS
Construct $\overline{PR} \cong \overline{HL}$.	Construct $\overline{PR} \cong \overline{HL}$.	Construct $\overline{PR} \cong \overline{HL}$.
Construct $\angle P \cong \angle H$.	Construct $\angle P \cong \angle H$.	Draw arc to begin constructing $\overline{PT} \cong \overline{HM}$.
Construct $\overline{PT} \cong \overline{HM}$.	Construct $\angle R \cong \angle L$.	Draw arc to begin constructing $\overline{RT} \cong \overline{LM}$.
Draw $\overline{TR}$.	Label point T.	Draw $\overline{PT}$ and $\overline{RT}$.

Using any method, △PRT ≅ △HLM.

Use SAS to complete the steps. Then construct a triangle congruent to △ABC. Check students' constructions.

1. Construct $\overline{PR} \cong$ __$\overline{AB}$__.

2. Construct $\angle P \cong$ __$\angle A$__.

3. Construct $\overline{PT} \cong$ __$\overline{AC}$__.

4. Draw __$\overline{TR}$__.

5. △PRT ≅ __△ABC__.

TAKE ANOTHER LOOK R49

Name _____

LESSON 12.1

Solid Figures

A *space figure*, or *solid figure*, is three-dimensional.
- The solid figure is a *polyhedron* if the sides *and* bases are polygons. A solid figure with a curved surface is *not* a polyhedron.
- The flat surfaces of a polyhedron are called *faces*.
- A sphere is not a polyhedron because it has a curved surface.
- A *prism* is a polyhedron with two congruent bases.
- The base tells you the name of the prism. Figure C below is a triangular prism because the base is a triangle.

Complete the table.

1.

	Solid Figure	Curved Surface?	Polyhedron?	Number of Faces	Shape of Base
A	 ⊖	yes	no	none	no base
B	▢	no	yes	6	square
C	△	no	yes	5	triangle
D	 ▯	yes	no	none	circle
E	 ◭	no	yes	5	square
F	 ⬡	no	yes	8	hexagon
G	 △	yes	no	none	circle
H	 ▭	no	yes	6	rectangle
I	 ⬠	no	yes	7	pentagon

Name each figure above. Tell if it is a polyhedron. If it is not, explain.

2. Figure A ___sphere; no; not a polyhedron___

3. Figure B ___cube; yes___

4. Figure D ___cylinder; no; not a polyhedron___

5. Figure E ___square pyramid; no; only one base___

6. Figure F ___hexagonal prism; yes___

7. Figure G ___cone; no; not a polyhedron___

8. Figure H ___rectangular prism; yes___

9. Figure I ___pentagonal prism; yes___

Name _____

LESSON 12.2

Problem-Solving Strategy

Finding Patterns in Polyhedrons

Polyhedrons have three parts:
- *Faces* are the flat sides and bases. They are polygons.
- *Edges* are segments where faces intersect.
- *Vertices* are points where edges intersect.

Faces, edges, and vertices are related.

To find how they are related, you can make a table of faces, vertices, and edges, and *find a pattern*.

Name of Figure	Number of Faces (F)	Number of Vertices (V)	Number of Edges (E)	F + V	F + V − E
Triangular prism	5	6	9	11	2
Rectangular prism	6	8	12	14	2
Triangular pyramid	4	4	6	8	2
Rectangular pyramid	5	5	8	10	2

You may find this pattern for prisms:
- As the number of faces increases by 1, the number of vertices increases by 2, and the number of edges increases by 3.

You may find this pattern for pyramids:
- As the number of faces increases by 1, the number of vertices increases by 1, and the number of edges increases by 2.

---

1. Is there a pattern for both prisms and pyramids? Complete the last two columns of the table above.

2. What is the pattern written as an equation? ___$F + V - E = 2$___

3. Solve the equation you found in Exercise 2 for E. ___$E = F + V - 2$___

For Exercises 4–6, use the pattern.

4. A pentagonal prism has 7 faces and 15 edges. How many vertices does it have?

    _____ $15 = 7 + V - 2$; $V = 10$; **10 vertices** _____

5. A hexagonal prism has 8 faces and 12 vertices. How many edges does it have?

    _____ $E = 8 + 12 - 2$; $E = 18$; **18 edges** _____

6. A prism has 20 vertices and 30 edges. How many faces does it have?

    _____ $30 = F + 20 - 2$; $F = 12$; **12 faces** _____

TAKE ANOTHER LOOK

Name _____

LESSON 12.3

Nets for Solid Figures

A *net* is a pattern for a solid figure. An arrangement of polygons can be folded to form a polyhedron.

A polyhedron can have different nets.

This net forms a cube.
- Let face 1 be on the bottom.
- Faces 2, 3, 4, and 5 are sides.
- Face 6 is on the top.

This net forms a square pyramid.
- Let face 1 be the base.
- Faces 2, 3, 4, and 5 are sides.

Complete.

1. This net forms a(n) __rectangular prism__.
   - Let face 1 be on the bottom.
   - Faces __2, 3, 4, and 5__ are sides.
   - Face __6__ is on the top.

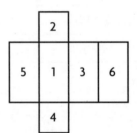

2. This net forms a(n) __cube__.
   - Let face C be on the bottom.
   - Faces __A, B, D, and F__ are sides.
   - Face __E__ is on the top.

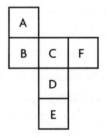

3. This net forms a __triangular prism__.
   - Let face O be on the bottom.
   - Faces __M, N, and P__ are sides.
   - Face __L__ is on the top.

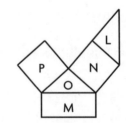

4. This net forms a __pentagonal pyramid__.
   - Face 1 is the __base__.
   - Faces __2, 3, 4, 5, and 6__ are sides.

R52 TAKE ANOTHER LOOK

Name _____

LESSON 12.4

Drawing Three-Dimensional Figures

Materials needed: straightedge, dot paper

Here is one way to draw a pyramid.

Step 1 Draw a base.

Step 2 Draw a point, P, above the base.

Step 3 Draw segments from point P to each vertex.

Step 4 Redraw "hidden edges" as dashed lines.

Here is one way to draw a prism.

Step 1 Draw a base.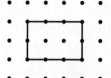

Step 2 Draw the other base above and to the right or left.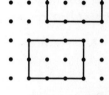

Step 3 Connect corresponding vertices to draw the lateral faces.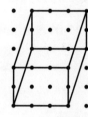

Step 4 Redraw "hidden edges" as dashed lines.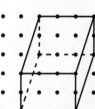

Draw the polyhedron. **Check students' drawings.**

1. square pyramid

2. cube

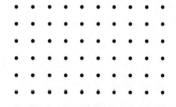

3. rectangular pyramid

4. pentagonal prism

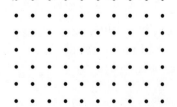

TAKE ANOTHER LOOK R53

Name _____

LESSON 13.1

Tessellations

A beehive is an example of a tessellation, where a figure is repeated over and over to make a pattern. The repeated figure in a beehive is a regular hexagon. The hexagons fit together so they cover the plane with no gaps or overlaps.

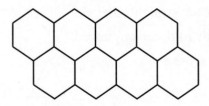

Squares and equilateral triangles are two other regular polygons that can be used to create tessellations. Make at least three rows of a tessellation using the given figure. **Check students' tessellations.**

1. a square

2. an equilateral triangle

You can change a polygon to create a new basic unit. Cut out a shape from one side of the polygon and move or translate that cut-out shape to the opposite side.

Step 1
Begin with a rectangle.

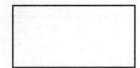

Step 2
Cut out a shape from one side.

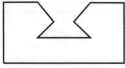

Step 3
Translate it to the opposite side.

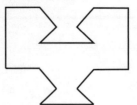

Step 4
Use the shape to make a tessellation.

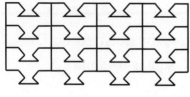

For Exercises 3 and 4, make the basic unit. Then use the basic unit to make at least two rows of a tessellation. **Check students' tessellations.**

3. Begin with a square. Cut out a semicircle from one side and translate it to the opposite side.

4. Begin with a hexagon. Cut out a triangle from one side and translate it to the opposite side.

R54 TAKE ANOTHER LOOK

LESSON 13.2

Geometric Iterations

A rotation is a transformation that turns a figure about a fixed point. Each of the following figures rotates about point *A*.

An iteration rule is used to describe a repeated rotation. The rule for the figures below is "Rotate the figure 90° clockwise about the point of rotation."

Stage 0   Stage 1   Stage 2   Stage 3   Stage 4

point of rotation

Every figure returns to its original position after rotating 360°. To find the number of stages necessary to return a figure to its original position, divide 360° by the measurement of each rotation in degrees.

So, the figure above will return to its original position after $\frac{360°}{90°}$, or 4 stages.

Rotate the figure 90° clockwise about the point of rotation. Complete the iteration process four times. Draw the figure at each stage. **Check students' drawings.**

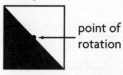

1. Stage 1   2. Stage 2   3. Stage 3   4. Stage 4

Rotate the figure 60° clockwise. Complete the iteration process four times. Draw the figure at each stage. **Check students' drawings.**

5. Stage 1   6. Stage 2   7. Stage 3   8. Stage 4

9. How many stages will it take the figure used in Exercises 5–8 to return to its original position? _____**6 stages**_____

TAKE ANOTHER LOOK

Name _____

LESSON 13.3

Self-Similarity

Study each of these three stages.

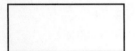 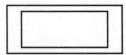 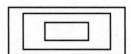

Stage 0
Rectangle

Stage 1
Rectangle in rectangle
The second is the same shape as the first but smaller.

Stage 2
Rectangle in rectangle, repeating
The third is the same shape as the first and second but smaller.

These stages show self-similarity. The smaller parts are like the whole but different in size.

Now study these stages.

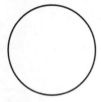   

Stage 0
Circle

Stage 1
Triangle in circle

Stage 2
Circle in triangle in circle

Stage 3
Triangle in circle, repeating

These stages do not show self-similarity because the shapes are different.

For Exercises 1–4, determine if the figure at Stage 2 has self-similarity. Then draw Stage 3 and determine if that stage has self-similarity. Explain.
Check students' drawings.

1.   Stage 3   __no, different shape;__

   Stage 0   Stage 1   Stage 2   __no, different shape__

2.   Stage 3   __yes, same shape;__

   Stage 0   Stage 1   Stage 2   __yes, same shape__

3.   Stage 3   __yes, same shape;__

   Stage 0   Stage 1   Stage 2   __yes, same shape__

4.   (Stage 2) Stage 3   __no, different shape;__

   Stage 0   Stage 1   Stage 2   __no, different shape__

R56 TAKE ANOTHER LOOK

Name _____

LESSON 13.4

Fractals

You can build a fractal by repeating a pattern. The pattern is a repetition of a shape in smaller and smaller sizes.

Stage 0

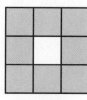
Stage 1

Stage 2

This fractal uses the following iteration process.

For Exercises 1–3, use the figures shown above.

1. How many shaded squares appear at Stage 0? at Stage 1? at Stage 2?

    __1; 8; 64__

2. Look for a pattern in the answers to Exercise 1. If this pattern continues, at Stage 3 there would be $8^{\boxed{3}}$, or ___512___, squares shaded.

3. At Stage n there would be $8^{\boxed{n}}$ squares shaded.

For Exercises 4 and 5, use the iteration rule below.

| Draw an equilateral triangle. | → | Reduce to $\frac{1}{2}$ size. Copy 3 times. Rebuild at each vertex. |

4. Draw Stage 2. Reduce to $\frac{1}{2}$ size. Copy 3 times. Rebuild at each vertex. **Check students' drawings.**

Stage 0

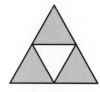
Stage 1

Stage 2

5. How many shaded triangles appear at Stage 0? at Stage 1? at Stage 2?

    at Stage n? ___1; 3; 9; 3^n___

TAKE ANOTHER LOOK R57

Name _____

LESSON 14.1

Problem-Solving Strategy

Drawing a Diagram to Show Ratios

Mike's Garage recommends changing the oil in your car every 3,000 mi and rotating the tires every 5,000 mi. You just bought a new car. How far will you drive before Mike's Garage changes the oil and rotates the tires at the same time?

The ratio of miles before an oil change to miles before rotating tires is 3:5.

Draw a diagram.

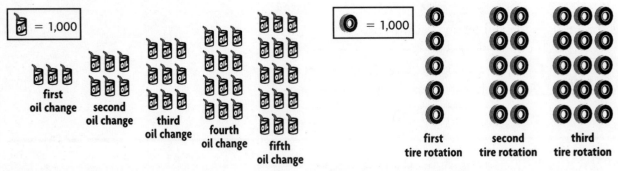

The diagram shows that the number of miles for an oil change equals the number of miles for rotating the tires at 15,000 mi.

Mike's Garage will change the oil and rotate the tires during the same visit at 15,000 mi.

Use the diagram at the right to solve.

1. Mrs. Snyder volunteered to bake cookies for a class party. The ratio of sugar to flour in the recipe is 1 to 4. How many cups of flour will she need if she uses

   3 c of sugar? _____12 c of flour_____

Sugar   Flour

Draw a diagram and solve.

2. The ratio of peanuts to raisins in the trail mix Maurice makes for his camping trip is 1 to 5. How many cups of raisins does he mix with

   4 c of peanuts? _____20 cups_____

3. The ratio of boys to girls in Charles's seventh-grade math class is 3:2. There are 15 boys in the class. How many girls are in the class?

   _____10 girls_____

4. The students at Mill Middle School can choose chocolate milk or plain milk for lunch. The ratio of students who choose chocolate milk to students who choose plain milk is 2 to 5. If 8 students select chocolate milk,

   how many select plain milk? _____20 students_____

R58   TAKE ANOTHER LOOK

LESSON 14.2

Ratios and Rates

Ted bicycles 86 km in 2 hr. What is his unit rate? The unit rate is the distance he bicycles in one unit of time, or 1 hr. Find the unit rate for 86 km in 2 hr.

Step 1 Write the rate as a fraction. $\quad\dfrac{86 \text{ km}}{2 \text{ hr}}$

Step 2 Divide the numerator *and* denominator by the denominator. $\quad\dfrac{86 \div 2}{2 \div 2} = \dfrac{43 \text{ km}}{1 \text{ hr}}$

Step 3 Write the fraction $\dfrac{43 \text{ km}}{1 \text{ hr}}$, using *per* to show division. $\quad\dfrac{43 \text{ km}}{1 \text{ hr}} = 43$ km per hr

A unit price is the cost for one unit of an item. Unit prices can help you choose a better buy.

Mrs. Chu needs to buy baby cereal. Which is a better buy, a 10-oz box of cereal for $1.79 or a 16-oz box for $2.55?

Step 1 Write each rate as a fraction.

Small Box $\quad\dfrac{\$1.79}{10 \text{ oz}}\quad$ Large Box $\dfrac{\$2.55}{16 \text{ oz}}$

Step 2 Divide so each denominator is 1.

$\dfrac{\$1.79 \div 10}{10 \text{ oz} \div 10} = \dfrac{\$0.18}{1 \text{ oz}} \qquad \dfrac{\$2.55 \div 16}{16 \text{ oz} \div 16} = \dfrac{\$0.16}{1 \text{ oz}}$

Step 3 Compare. $\quad$ $0.18 per oz > $0.16 per oz, so the 16-oz box for $2.55 is the better buy.

Complete to find the unit rate.

1. $68 in 16 hr

   Step 1 $\dfrac{\$68}{16 \text{ hr}}$

   Step 2 $\dfrac{\$68 \div \boxed{16}}{16 \text{ hr} \div \boxed{16}} = \dfrac{\boxed{\$4.25}}{1 \text{ hr}}$

   Step 3 $\dfrac{\boxed{\$4.25}}{1 \text{ hr}} = \boxed{\$4.25}$ per hr

2. 196 km in 7 hr

   Step 1 $\dfrac{196 \text{ km}}{7 \text{ hr}}$

   Step 2 $\dfrac{196 \text{ km} \div \boxed{7}}{7 \text{ hr} \div \boxed{7}} = \dfrac{\boxed{28} \text{ km}}{1 \text{ hr}}$

   Step 3 $\dfrac{\boxed{28 \text{ km}}}{\boxed{1 \text{ hr}}} = \boxed{28 \text{ km}}$ per hr

Which is the better buy?

3. an 8-oz container of yogurt for $0.69 or a 32-oz container for $2.49

   The __32-oz__ container is the better buy.

4. a 10-oz bag of trail mix for $0.89 or a 16-oz bag for $1.49

   The __10-oz__ bag is the better buy.

TAKE ANOTHER LOOK   R59

Name _____

LESSON 14.3

Rates in Tables and Graphs

Tables and graphs can help you organize data. This can make predicting rates easy.

What is the pattern in each table below?

Distance	Gallons of Gas Used
100 mi	4 gal
200 mi	8 gal
400 mi	16 gal
800 mi	32 gal

× 2, × 2, × 2 (left) ; × 2, × 2, × 2 (right)

Weight	Cost
256 lb	$192
64 lb	$48
16 lb	$12
4 lb	$3

÷ 4, ÷ 4, ÷ 4 (left) ; ÷ 4, ÷ 4, ÷ 4 (right)

Pattern: Each value is twice the number above it.

Use the pattern you found above to predict how many gallons you would need to drive 1,600 mi.

- 800 × 2 = 1,600, and 32 × 2 = 64.
- You would need 64 gal.

Pattern: Each value is one-fourth the number above it.

Use the pattern you found above to predict the cost of 1 lb.

- 4 ÷ 4 = 1, and 3 ÷ 4 = 0.75.
- The cost would be $0.75.

Complete.

1.

Number of Tickets	Cost
5	$45
15	$135
45	$405
135	$1,215

What is the pattern?

__Each value is three times the number above it.__

Find the cost of 405 tickets.

135 × __3__ = __405__,

and 1,215 × __3__ = __3,645__.

The cost of 405 tickets is __$3,645__.

2.

Distance	Time
162 m	64.8 sec
54 m	21.6 sec
18 m	7.2 sec
6 m	2.4 sec

What is the pattern?

__Each value is one-third the number above it.__

Find the time to travel 2 m.

6 ÷ __3__ = __2__,

and 2.4 ÷ __3__ = __0.8__.

The time to travel 2 m is __0.8 sec__.

R60 TAKE ANOTHER LOOK

Name _____

LESSON 14.4

Finding Golden Ratios

A Golden Cut divides a segment into two segments with the following ratios:
- $\frac{\text{longer segment}}{\text{shorter segment}} \approx 1.61$
- $\frac{\text{total segment}}{\text{longer segment}} \approx 1.61$
- Remember, 1.61 is the *Golden Ratio*.

Point *B* divides segment *AC*. Determine if point *B* makes a Golden Cut, also known as a Golden Section.

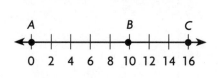

$\overline{AB} = 9.9$ Find the two ratios:
$\overline{BC} = 6.1$ $\frac{\text{longer segment}}{\text{shorter segment}} = \frac{AB}{BC} = \frac{9.9}{6.1} \approx 1.6$
$\overline{AC} = 16$ $\frac{\text{total segment}}{\text{longer segment}} = \frac{AC}{AB} = \frac{16}{9.9} \approx 1.6$

Both ratios are approximately equal to 1.61, so point *B* makes a Golden Cut of segment *AC*.

There is a helpful shortcut. If one ratio is approximately equal to 1.61, then the other one is also. So, you need to find only one ratio to determine if a cut is a Golden Cut.

For Exercises 1–3, complete to determine if point *Q* makes a Golden Cut of segment *PR*.

1.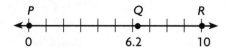

   $PQ = 6.2$

   $QR =$ __3.8__

   $PR =$ __10__

   $\frac{\text{longer segment}}{\text{shorter segment}} = \frac{PQ}{QR} = \boxed{\frac{6.2}{3.8}} \approx$ __1.63__

   Does point *Q* make a Golden Cut? __yes__

2.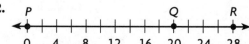

   $PQ =$ __20__

   $QR =$ __8__

   $PR =$ __28__

   $\frac{\text{longer segment}}{\text{shorter segment}} = \boxed{\frac{PQ}{QR}} = \boxed{\frac{20}{8}} \approx$ __2.5__

   Does point *Q* make a Golden Cut? __no__

3.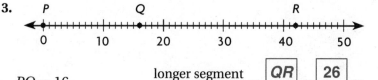

   $PQ = 16$

   $QR =$ __26__

   $PR = 42$

   $\frac{\text{longer segment}}{\text{shorter segment}} = \boxed{\frac{QR}{PQ}} = \boxed{\frac{26}{16}} \approx$ __1.63__

   Does point *Q* make a Golden Cut? __yes__

TAKE ANOTHER LOOK

Name _____

LESSON 15.1

Changing Ratios to Percents

You can represent the ratio 3 to 4 as 3:4 or $\frac{3}{4}$.

You have studied two ways to write the ratio $\frac{3}{4}$ as a percent:

- Write an equivalent ratio with a denominator of 100. $\frac{3}{4} = \frac{75}{100} = 75\%$

- Use a proportion to change a ratio to a percent. $\frac{3}{4} = \frac{x}{100}$, $x = 75$, $\frac{3}{4} = 75\%$

Now you will learn a third way, using division.

Use division to write the ratios as percents.

a. $\frac{3}{4}$ b. $\frac{12}{8}$

$3 \div 4 = 0.75$ $12 \div 8 = 1.5$ *Divide the numerator by the denominator.*

$0.75 \times 100 = 75$ $1.5 \times 100 = 150$ *Multiply by 100 to change the decimal to a percent.*

$\frac{3}{4}$ is 75%. $\frac{12}{8}$ is 150%.   *Add a percent symbol.*

Match the ratio to the equivalent percent.

1. $\frac{9}{12}$ __e__   a. 12.5%   2. 12 to 3 __b__   a. 80%

    7:25 __c__   b. 5% $\frac{8}{5}$ __d__   b. 400%

    2 to 16 __a__   c. 28%   3:10 __e__   c. 35%

    1:20 __b__   d. 125%   4 to 5 __a__   d. 160%

    $\frac{5}{4}$ __d__   e. 75% $\frac{7}{20}$ __c__   e. 30%

Write each ratio as a percent.

3. $\frac{6}{8}$ __75%__   4. 9 to 500 __1.8%__   5. 10 to 8 __125%__   6. 3:100 __3%__

7. $\frac{3}{5}$ __60%__   8. 1:50 __2%__   9. 1 to 8 __12.5%__   10. $\frac{1}{4}$ __25%__

11. $\frac{9}{2}$ __450%__   12. 4:20 __20%__   13. $\frac{7}{8}$ __87.5%__   14. 4:5 __80%__

15. 16 to 10 __160%__   16. 19 to 20 __95%__   17. $\frac{18}{24}$ __75%__   18. 75:50 __150%__

19. $\frac{6}{15}$ __40%__   20. 12:300 __4%__   21. 3 to 8 __37.5%__   22. $\frac{9}{8}$ __112.5%__

23. $\frac{7}{4}$ __175%__   24. 2:25 __8%__   25. $\frac{5}{8}$ __62.5%__   26. 2:5 __40%__

TAKE ANOTHER LOOK

LESSON 15.2

Finding a Percent of a Number

To find a percent of a number, you can change the percent to a fraction or a decimal.

Percent	Fraction	Decimal
	Drop the percent sign, and use 100 as the denominator.	Move the decimal point two places to the left, and drop the percent sign.
14%	$\frac{14}{100} = \frac{7}{50}$	14% = 0.14
300%	$\frac{300}{100} = 3$	300% = 3

Find 6% of 75.

	Fraction	Decimal
Change the percent to a fraction or a decimal.	$\frac{6}{100}$ of 75	0.06 of 75
Simplify the fraction, if possible.	$\frac{3}{50}$ of 75	
Replace the word *of* with the multiplication symbol.	$\frac{3}{50} \times 75$	0.06×75
Multiply.	$\frac{3}{50} \times \frac{75}{1} = \frac{9}{2} = 4\frac{1}{2}$	4.5

So, 6% of 75 = $4\frac{1}{2}$, or 4.5.

Find the percent of each number.

1. 50% of 32 __16__
2. 100% of 75 __75__
3. 90% of 18 __16.2__
4. 140% of 240 __336__
5. 5% of 38 __1.9__
6. 25% of 120 __30__
7. 80% of 160 __128__
8. 75% of 16 __12__
9. 0.2% of 68 __0.136__
10. 250% of 80 __200__
11. 55% of 52 __28.6__
12. 7% of 3,200 __224__
13. 10% of 96 __9.6__
14. 9% of 960 __86.4__
15. 39% of 320 __124.8__
16. 6.25% of 500 __31.25__
17. 105% of 840 __882__
18. 2% of 16 __0.32__
19. 0.3% of 20 __0.06__
20. 125% of 44 __55__
21. 14% of 350 __49__

TAKE ANOTHER LOOK   R63

Name _____

LESSON 15.3

Finding What Percent One Number Is of Another

To find what percent one number is of another, do the following:

What percent of 15 is 6?
           ↑ ↑
     whole part

$\frac{part}{whole} \rightarrow \frac{6}{15} = \frac{2}{5}$ *Write the ratio of part to whole as a fraction in simplest form.*

$5\overline{)2.00}^{\,0.40} = 40\%$ *Divide to write the ratio as a percent.*

So, 40% of 15 is 6.

Complete the table.

	Question	Ratio	Percent
1.	What percent of 100 is 19?	$\frac{19}{100}$	19%
2.	What percent of 200 is 78?	$\frac{78}{200} = \frac{39}{100}$	39%
3.	What percent of 8 is 4?	$\frac{4}{8} = \frac{1}{2}$	50%
4.	What percent of 200 is 280?	$\frac{280}{200} = \frac{7}{5}$	140%
5.	What percent of 30 is 6?	$\frac{6}{30} = \frac{1}{5}$	20%
6.	What percent of 36 is 9?	$\frac{9}{36} = \frac{1}{4}$	25%
7.	What percent of 1,000 is 60?	$\frac{60}{1,000} = \frac{3}{50}$	6%
8.	What percent of 12 is 24?	$\frac{24}{12} = 2$	200%
9.	What percent of 20 is 8?	$\frac{8}{20} = \frac{2}{5}$	40%
10.	What percent of 96 is 36?	$\frac{36}{96} = \frac{3}{8}$	37.5%
11.	What percent of 15 is 3?	$\frac{3}{15} = \frac{1}{5}$	20%
12.	What percent of 100 is 125?	$\frac{125}{100} = \frac{5}{4}$	125%
13.	What percent of 200 is 720?	$\frac{720}{200} = \frac{18}{5}$	360%
14.	What percent of 240 is 72?	$\frac{72}{240} = \frac{3}{10}$	30%
15.	What percent of 110 is 11?	$\frac{11}{110} = \frac{1}{10}$	10%
16.	What percent of 93 is 31?	$\frac{31}{93} = \frac{1}{3}$	$33\frac{1}{3}\%$

R64 TAKE ANOTHER LOOK

Name _____

LESSON 15.4

Finding a Number When the Percent Is Known

The same question can be asked in different ways:
- 95 is 25% of what number?
- 25% of what number is 95?

To answer the question, you can follow three steps:

Step 1 Change the percent to a decimal.   25% = 0.25

Step 2 Write the ratio $\frac{\text{given number}}{\text{decimal}}$. $\frac{95}{0.25}$

Step 3 Divide.   95 ÷ 0.25 = 380.

So, 95 is 25% of 380.

Complete the table.

	Question	Ratio	Percent
1.	25% of what number is 30?	$\frac{30}{0.25}$	120
2.	18 is 6% of what number?	$\frac{18}{0.06}$	300
3.	16% of what number is 19.2?	$\frac{19.2}{0.16}$	120
4.	250% of what number is 225?	$\frac{225}{2.5}$	90
5.	99.6 is 60% of what number?	$\frac{99.6}{0.6}$	166
6.	48 is 8% of what number?	$\frac{48}{0.08}$	600
7.	40% of what number is 3.84?	$\frac{3.84}{0.4}$	9.6
8.	64 is 50% of what number?	$\frac{64}{0.50}$	128
9.	75% of what number is 9.45?	$\frac{9.45}{0.75}$	12.6
10.	125 is 100% of what number?	$\frac{125}{1.0}$	125
11.	90% of what number is 18?	$\frac{18}{0.90}$	20
12.	13 is 10% of what number?	$\frac{13}{0.10}$	130
13.	80% of what number is 128?	$\frac{128}{0.80}$	160
14.	5% of what number is 48?	$\frac{48}{0.05}$	960

TAKE ANOTHER LOOK

Lesson 16.1

Similar Figures and Scale Factors

When you enlarge a figure, as with an overhead projector, you get a figure that is similar to the original figure.

Similar Figures
- ✓ Congruent Angles
- ✓ Proportional Sides

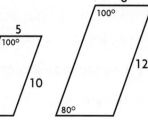

Decide if these figures are similar.

Step 1 Check for congruent angles. Yes; corresponding angles are congruent.

Step 2 Check for proportional sides: $\frac{12}{10} = \frac{6}{5}$. Yes; corresponding sides are proportional.

Conclusion: The figures are similar.

Determine whether the figures are similar. Write *yes* or *no*, and support your answer.

1.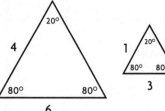

   no; $\frac{4}{1} \neq \frac{6}{3}$

2.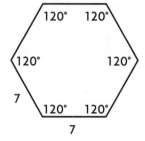

   yes; angles congruent and $\frac{7}{2} = \frac{7}{2}$

Write *true* or *false*.

3. Similar figures are always the same shape and the same size.   **false**

4. A proportion is an equation stating that two ratios are equivalent.   **true**

5. The common ratio for pairs of corresponding sides of similar figures is called the scale factor.   **true**

6. Similar figures always have congruent pairs of corresponding sides.   **false**

7. The sides of $\triangle DEF$ are 6 in., 8 in., and 10 in. A smaller, similar triangle constructed with a scale factor of $\frac{1}{4}$ would have sides of length 1.5 in., 2 in., and 2.5 in.   **true**

R66 TAKE ANOTHER LOOK

Name _____

LESSON 16.2

Proportions and Similar Figures

When two figures are similar, you can use a proportion to solve for an unknown length of a side.

Look at the two figures. Which proportion would you use to solve for x? Why?

a. $\frac{8}{x} = \frac{2}{12}$ **b.** $\frac{8}{12} = \frac{2}{x}$

You should use proportion b because the ratios are made up of *corresponding sides*.

Think: $\frac{\text{left side}}{\text{left side}} = \frac{\text{top}}{\text{top}}$

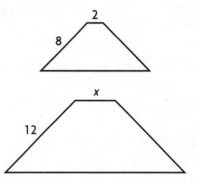

Find the unknown length in the similar figures.

Step 1 Write a proportion. $\frac{10}{20} = \frac{7.5}{x}$

Step 2 Find the cross products. $10x = 20 \times 7.5$

Step 3 Solve. $10x = 150;\ x = 15$

The unknown length is 15 in.

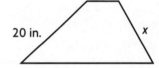

Answer the following, using the similar triangles shown.

1. Which proportion should you use to solve for x?

   a. $\frac{4}{x} = \frac{8}{10}$ **b.** $\frac{4}{10} = \frac{8}{x}$

   _____a_____

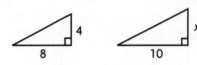

2. Name the pairs of corresponding sides in triangles *BAT* and *PIN*.

   _____*BA* and *PI*, *AT* and *IN*, *TB* and *NP*_____

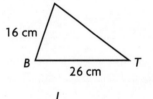

3. Complete the proportion to find n.

   $\dfrac{\boxed{16}}{\boxed{24}} = \dfrac{\boxed{26}}{\boxed{n}}$

   n is ____39____ cm.

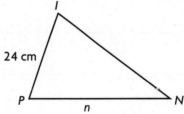

4. Use a proportion to find the unknown length in the similar figures. $x =$ ____12.5____

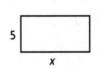
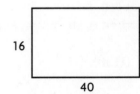

TAKE ANOTHER LOOK R67

Name _____

LESSON 16.3

Areas of Similar Figures

Cut out a 4-in. × 4-in. piece of paper and an 8-in. × 8-in. piece of paper. How many of the smaller squares does it take to cover the larger square?

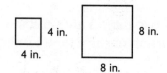

It takes four. The sides of the larger square are *twice* as great, but the area of the larger square is *four* times as great.

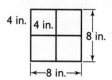

ratio, or scale factor of sides: $\frac{8}{4} = \frac{2}{1}$

ratio, or scale factor of areas: $\frac{64}{16} = \frac{4}{1}$ or $\boxed{2^2}$

Now compare a 3-cm × 3-cm square to a 9-cm × 9-cm square. The sides are *three* times as great, but the area is *nine* times as great.

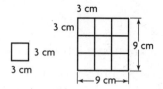

ratio, or scale factor of sides: $\frac{9}{3} = \frac{3}{1}$

ratio, or scale factor of areas: $\frac{81}{9} = \frac{9}{1}$ or $\boxed{3^2}$

In general, the scale factor for area is the *square* of the scale factor for the sides of the figures.

Complete the table.

	Scale Factor for the Sides of Two Similar Polygons	Scale Factor for Their Areas	Relationship of Scale Factors
	7:5	$7^2:5^2$, 49:25, or $\frac{49}{25}$	$\left(\frac{7}{5}\right)^2 = \frac{49}{25}$
1.	$\frac{10}{3}$	100:9, or $\frac{100}{9}$	$\left(\frac{10}{3}\right)^2 = \frac{100}{9}$
2.	6:1	36:1, or $\frac{36}{1} = 36$	$\left(\frac{6}{1}\right)^2 = \frac{36}{1}$, or 36
3.	9:2	81:4, or $\frac{81}{4}$	$\left(\frac{9}{2}\right)^2 = \frac{81}{4}$
4.	$\frac{11}{7}$	121:49, or $\frac{121}{49}$	$\left(\frac{11}{7}\right)^2 = \frac{121}{49}$

For Exercises 5–6, use the similar triangles.

5. Find the area of the original triangle and then of the enlargement.

    (Hint: $A = \frac{1}{2}bh$) ___30 in.²; 67.5 in.²___

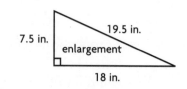

6. Find the scale factor for the sides and then for the areas. ___$\frac{3}{2}$ or 3:2; $\frac{9}{4}$ or 9:4___

R68   TAKE ANOTHER LOOK

Volumes of Similar Figures

LESSON 16.4

What is the relationship between the volumes of similar figures and their dimensions?

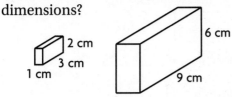

volume of a prism = length × width × height

$V = l \times w \times h$
$= 1 \times 3 \times 2$
$= 6; \ 6 \text{ cm}^3$

$V = l \times w \times h$
$= 3 \times 9 \times 6$
$= 162; \ 162 \text{ cm}^3$

ratio, or scale factor, for volumes:
$\frac{162}{6} = \frac{27}{1}$ or $\boxed{3^3}$

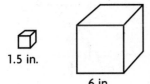

$V = 1.5 \times 1.5 \times 1.5$
$= 3.375; \ 3.375 \text{ in.}^3$

$V = 6 \times 6 \times 6$
$= 216; \ 216 \text{ in.}^3$

ratio of volumes:
$\frac{216}{3.375} = \frac{64}{1} = \boxed{4^3}$

Look for a relationship between the scale factor for the volumes and the scale factor for the sides. Notice that the ratio of the volumes is the *cube* of the scale factor for the sides.

Complete the table.

	Dimension of Original Cube	Dimension of Enlarged Cube	Scale Factor	Scale Factor of Volumes
1.	2 in.	3 in.	$\frac{3}{2}$, or 1.5	$\frac{27}{8}$, or 3.375
2.	4 cm	12 cm	3	27
3.	4.5 ft	9 ft	2	8

Find the volumes of the similar prisms. Then write the scale factor for the volumes.

4.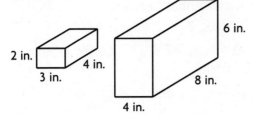

5.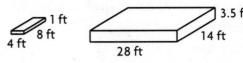

$V_{small} = 24 \text{ in.}^3; \ V_{large} = 192 \text{ in.}^3; \ 8$

$V_{small} = 32 \text{ ft}^3; \ V_{large} = 1{,}372 \text{ ft}^3; \ 42.875$

TAKE ANOTHER LOOK R69

Name _____

LESSON 17.1

Drawing Similar Figures

Use the scale factor to draw a similar figure. Remember:
- If the scale factor is greater than 1, the new figure is larger.
- If the scale factor is less than 1, the new figure is smaller.

To draw a similar figure, be sure the corresponding angles have the same measures. Be sure each corresponding side is multiplied by a scale factor.

Using a scale factor of 1.5, draw a larger triangle that is similar to the triangle above.

$1.5 \times 6 = 9$
$1.5 \times 8 = 12$
$1.5 \times 10 = 15$

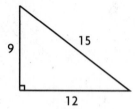

Now use a scale factor of 0.5 to draw a smaller similar triangle.

$0.5 \times 6 = 3$
$0.5 \times 8 = 4$
$0.5 \times 10 = 5$

Suppose you are asked to draw a trapezoid similar to the trapezoid below.

1. If the scale factor is $\frac{1}{6}$, the new figure will be ___smaller___ than the original trapezoid.

2. With a scale factor of $\frac{1}{6}$, the lengths of the sides of the new trapezoid will be

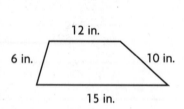

a. $\frac{1}{6} \times$ __6__ = __1__ in.

b. $\frac{1}{6} \times$ __12__ = __2__ in.

c. $\frac{1}{6} \times$ __10__ = $1\frac{2}{3}$ in.

d. $\frac{1}{6} \times$ __15__ = $2\frac{1}{2}$ in.

Suppose you are asked to draw a triangle similar to the triangle below. Use a scale factor of 2.25.

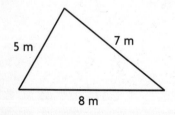

3. A scale factor of 2.25 will ___enlarge___ the size of the original triangle.

4. The lengths of the sides of the new triangle will be

___11.25___ m, ___15.75___ m, and ___18___ m.

Name _____

LESSON 17.2

Scale Drawings

At the right is a scale drawing of a guest room. The scale is 1 in.: 8 in. This means that 1 in. in the scale drawing represents an actual length of 8 in. in the real room.

The ratio is $\frac{1 \text{ in.}}{8 \text{ in.}}$ ← length in drawing ← actual length

If you know the length in the drawing, you can use this ratio to find the actual length.

For example, you can use a proportion to find the actual dimensions of the guest room. Let n represent the actual length.

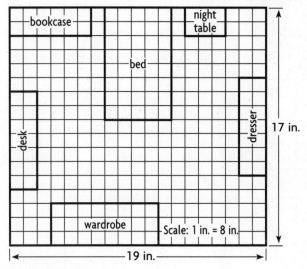

	Length of Room	Width of Room
Write a proportion.	$\frac{\text{length in drawing}}{\text{actual length}} : \frac{1 \text{ in.}}{8 \text{ in.}} = \frac{19 \text{ in.}}{n \text{ in.}}$	$\frac{1 \text{ in.}}{8 \text{ in.}} = \frac{17 \text{ in.}}{n \text{ in.}}$
Cross multiply.	$1 \times n = 8 \times 19$	$1 \times n = 8 \times 17$
Solve for n.	$n = 152$ in.	$n = 136$ in.

The actual dimensions of the guest room are 152 in. by 136 in.

Find the dimensions of objects in the guest room.

1. Find the length of the bed.

   $\frac{1 \text{ in.}}{8 \text{ in.}} = \frac{8 \text{ in.}}{n \text{ in.}}$

   $1 \times n = \underline{8 \times 8}$

   $n = \underline{64}$ in.

2. Find the width of the bed.

   $\frac{1 \text{ in.}}{8 \text{ in.}} = \frac{\boxed{5 \text{ in.}}}{n \text{ in.}}$

   $1 \times n = \underline{8 \times 5}$

   $n = \underline{40}$ in.

3. The actual length of the wardrobe is __64__ in.

4. The actual width of the bookcase is __16__ in.

5. The dimensions of the dresser are __16__ in. × __56__ in.

6. The dimensions of the desk are __16__ in. × __56__ in.

7. The dimensions of the night table are __16__ in. × __24__ in.

8. If the actual dimensions of a room are 96 in. × 120 in., a scale drawing of the room would be __12__ in. × __15__ in.

TAKE ANOTHER LOOK

Name _____

LESSON 17.3

Using Maps

Materials needed: metric ruler

How can you find the straight-line distance between Lakefield and Ashton?

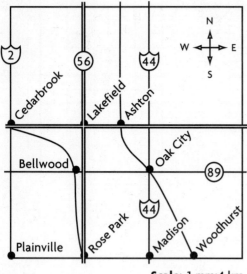

Scale: 1 mm : 4 km

Step 1 Measure the distance in mm between the two cities on the map. The distance between Lakefield and Ashton is 10 mm on the map.

Step 2 Write the scale as a ratio.

$\frac{1 \text{ mm}}{4 \text{ km}}$ ← map distance
← actual distance

Step 3 Write a proportion to find the actual distance from Lakefield to Ashton.

$\frac{1 \text{ mm}}{4 \text{ km}} = \frac{10 \text{ mm}}{n \text{ km}}$

Step 4 Solve the proportion.

$1 \times n = 4 \times 10$
$n = 40$

The distance between Lakefield and Ashton is 40 km.

Use the map above to complete the table.

	Cities	Distance on Map (in mm)	Proportion	Actual Distance (in km)
1.	Cedarbrook to Ashton	30 mm	$\frac{1 \text{ mm}}{4 \text{ km}} = \frac{\boxed{30} \text{ mm}}{n \text{ km}}$	120 km
2.	Rose Park to Lakefield	35 mm	$\frac{1 \text{ mm}}{4 \text{ km}} = \frac{35 \text{ mm}}{n \text{ km}}$	140 km
3.	Plainville to Woodhurst	50 mm	$\frac{1 \text{ mm}}{4 \text{ km}} = \frac{50 \text{ mm}}{n \text{ km}}$	200 km
4.	Bellwood to Oak City	20 mm	$\frac{1 \text{ mm}}{4 \text{ km}} = \frac{20 \text{ mm}}{n \text{ km}}$	80 km
5.	Madison to Plainville	38 mm	$\frac{1 \text{ mm}}{4 \text{ km}} = \frac{38 \text{ mm}}{n \text{ km}}$	152 km
6.	Cedarbrook to Lakefield	20 mm	$\frac{1 \text{ mm}}{4 \text{ km}} = \frac{20 \text{ mm}}{n \text{ km}}$	80 km
7.	Madison to Woodhurst	12 mm	$\frac{1 \text{ mm}}{4 \text{ km}} = \frac{12 \text{ mm}}{n \text{ km}}$	48 km

TAKE ANOTHER LOOK

Name _____

LESSON 17.4

Indirect Measurement

Marcy wants to know the height of a cactus. Marcy is 50 in. tall. Her shadow is 30 in. long, and the shadow of the cactus is 76 in. long. Use similar triangles to measure indirectly.

Step 1
Draw and label two similar triangles.

Step 2
Write and solve a proportion.

Marcy's shadow → $\frac{30}{76} = \frac{50}{n}$ ← Marcy's height
cactus's shadow →   ← cactus's height

$30n = 50 \times 76$

$30n = 3{,}800$

$\frac{30n}{30} = \frac{3{,}800}{30}$

$n = 126\frac{2}{3}$

So, the height of the cactus is $126\frac{2}{3}$ in.

The triangles in each pair are similar. Fill in the boxes to find the length of *x*.

1.
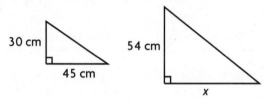

$\frac{30}{54} = \frac{45}{x}$

$\boxed{30}\, x = \boxed{45} \times 54$

$x = \underline{81\text{ cm}}$

2.
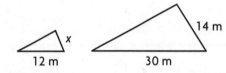

$\frac{12}{30} = \dfrac{x}{\boxed{14}}$

$12 \times \boxed{14} = \boxed{x} \times 30$

$x = \underline{5.6\text{ m}}$

3. Draw and label two similar triangles. Then write a proportion and solve. **Check students' drawings, proportions, and solutions.**

4. Anne and her brother Sam are standing next to each other. Anne casts a 90-cm shadow and Sam casts a 108-cm shadow. If Sam is 162 cm tall, how tall is Anne? Anne is __135__ cm tall.

TAKE ANOTHER LOOK R73

Name _____

LESSON 17.5

Golden Rectangles

A Golden Rectangle is a visually pleasing rectangle with a length-to-width ratio of about 1.61:1, or 1.61.

If you cut a square from one end of a Golden Rectangle, the rectangle that remains is also a Golden Rectangle. Find the ratio for the new Golden Rectangle.
$\frac{\text{length}}{\text{width}} = \frac{1}{0.61} = 1.639$

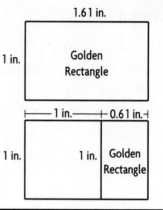

---

1. Find the ratio $\frac{\text{length}}{\text{width}}$ for the rectangle below.

   26 mm ▭ 42 mm

   $\frac{42}{26} = \underline{\ 1.62\ }$

2. Is the rectangle a Golden Rectangle? __yes__

3. Suppose you cut a square from the rectangle in Exercise 1. Draw the new rectangle. **Check students' drawings. The rectangle should be 26 mm by 16 mm.**

4. Find the ratio $\frac{\text{length}}{\text{width}}$ for the new rectangle. $\frac{26}{16} = \underline{\ 1.625\ }$

5. Is the new rectangle a Golden Rectangle? __yes__

6. Suppose you cut a square from the rectangle in Exercise 3. Draw the new rectangle. **Check students' drawings. The rectangle should be 16 mm by 10 mm.**

7. Is the new rectangle a Golden Rectangle? __yes__

Complete the table. Is each a Golden Rectangle?

	Dimensions	Ratio	Yes or No
8.	16 cm by 9 cm	$\frac{16}{9} \approx 1.778$	No
9.	102 mm by 63 mm	$\frac{102}{63} \approx 1.619$	Yes
10.	19 in. by 11.8 in.	$\frac{19}{11.8} \approx 1.61$	Yes
11.	50 cm by 40 cm	$\frac{50}{40} \approx 1.25$	No

R74 TAKE ANOTHER LOOK

Name _____

LESSON 18.1

Triangular Arrays

An array is an orderly group of numbers or shapes. When the group looks like a triangle, it is called a *triangular array*.

For example, the first five counting numbers are shown in a triangular array.

```
              1
            1   2
          1   2   3
        1   2   3   4
      1   2   3   4   5
```

The numbers formed by adding each row in the array are called *triangular numbers*.

To find a triangular number, you can use an array, a pattern, or a formula. What is the fourth triangular number? the fifth triangular number?

Triangular Number	Array	Pattern	Formula $\frac{n(n+1)}{2}$
Fourth	4 in bottom row, 10 circles	$1 + 2 + 3 + 4 = 10$	$\frac{4(4+1)}{2} = \frac{4(5)}{2} = 10$
Fifth	5 in bottom row, 15 circles	$1 + 2 + 3 + 4 + 5 = 15$	$\frac{5(5+1)}{2} = \frac{5(6)}{2} = 15$

The fourth triangular number is 10. The fifth triangular number is 15.

Complete the table to find each triangular number.

	Triangular Number	Array	Pattern	Formula $\frac{n(n+1)}{2}$
1.	Eighth	Check students' arrays.	$1 + 2 + 3 + 4 + 5 + 6 + 7 + 8 = \underline{36}$	$\frac{8(8+1)}{2} = \underline{36}$
2.	Tenth	Check students' arrays.	$1 + 2 + 3 + 4 + 5 + 6 + 7 + 8 + 9 + 10 = 55$	$\frac{10(10+1)}{2} = 55$

Use the formula to find the triangular number.

3. ninety-fourth $\underline{\frac{94(94+1)}{2}}$ = $\underline{4,465}$

TAKE ANOTHER LOOK R75

Name _____

LESSON 18.2

Pascal's Triangle

```
                                              Sum of Numbers in Row
Row 0:                  1                            1
Row 1:              1       1                      1 + 1 = 2
Row 2:           1     2       1                 1 + 2 + 1 = 4
Row 3:        1     3     3       1            1 + 3 + 3 + 1 = 8
Row 4:     1     4     6     4       1       1 + 4 + 6 + 4 + 1 = 16
```

Here is how to build Pascal's triangle:
- In each row of Pascal's triangle, the first and last entries are 1.
- Each of the other numbers is the sum of the two closest numbers above it.

---

1. Find the numbers in Row 5. __1, 5, 10, 10, 5, 1__

2. Row 4 has 5 numbers. Row 5 has 6 numbers. How many numbers are in Row 6? in Row 7? in Row 8? __7 numbers; 8 numbers; 9 numbers__

3. What is the sum of the numbers in Row 5?

   __1__ + __5__ + __10__ + __10__ + __5__ + __1__ = __32__

Hot Diggity Dog offers 3 toppings for hot dogs: ketchup, mustard, and relish. How many kinds of hot dogs can be made with no toppings, 1 topping, 2 toppings, or all 3 toppings?

Complete.

4. You have a choice of __3__ toppings, so look at Row __3__ of Pascal's triangle.

5. The first number in the row, __1__, represents 0 toppings.

6. The second number in the row, __3__, represents the ways you can choose 1 topping.

7. The __third__ number in the row, __3__, represents the ways you can choose 2 toppings.

8. The __fourth__ number in the row, __1__, represents the ways you can choose all __3__ toppings.

9. What is the number of possible hot dogs that can be made?

   __1__ + __3__ + __3__ + __1__ = __8__

R76 TAKE ANOTHER LOOK

Repeated Doubling and Halving

LESSON 18.3

What happens to a number when it is doubled or halved over and over again? Let's see what happens to the number 64.

Doubled	Halved
$64 \times 2 = 128$	$64 \times \frac{1}{2} = 32$
$128 \times 2 = 256$	$32 \times \frac{1}{2} = 16$
$256 \times 2 = 512$	$16 \times \frac{1}{2} = 8$
$512 \times 2 = 1{,}024$	$8 \times \frac{1}{2} = 4$

1. When you double a positive number, the doubles grow, or __diverge__.

2. When you halve a positive number, the halves shrink, or __converge__.

3. The next row of the table above should have $\boxed{1{,}024} \times 2 = \boxed{2{,}048}$ in the first column and $\boxed{4} \times \frac{1}{2} = \boxed{2}$ in the second column.

4. Use the table to help you find how many times 64 must be doubled to reach or exceed 1,000. How many times must 64 be halved to reach or be less than 10? __4 times; 3 times__

5. Begin with the number 7. Double 7 ten times. Fill in a box each time you double.

   7; $\boxed{14}$; $\boxed{28}$; $\boxed{56}$; 112; $\boxed{224}$; $\boxed{448}$; 896; $\boxed{1{,}792}$; 3,584; $\boxed{7{,}168}$

6. Begin with the number 3,000. Halve 3,000 six times. Fill in a box each time you halve.

   3,000; $\boxed{1{,}500}$; $\boxed{750}$; 375; 187.5; $\boxed{93.75}$; $\boxed{46.875}$

Roberta buys 4 guppies. Every month the guppy population doubles.

7. After 1 month there will be __4__ × 2 or __8__ guppies.

8. What will be Roberta's guppy population after 2 months? after 3 months? after 4 months? after 5 months? __16 guppies; 32 guppies; 64 guppies; 128 guppies__

Mrs. Diaz baked 48 muffins. Every day she will give away $\frac{1}{2}$ of the muffins that are left.

9. On the first day, Mrs. Diaz will give away __48__ $\times \frac{1}{2}$ or __24__ muffins.

10. How many muffins will Mrs. Diaz give away on the second day? the third day? the fourth day? __12 muffins, 6 muffins, 3 muffins__

TAKE ANOTHER LOOK R77

Name _____

LESSON 18.4

Exponents and Powers

An exponent tells how many times a base is used as a factor. Numbers that are expressed using exponents are called powers.

You can find powers of any number. The first power is the number itself. There are as many factors as the exponent shows.

$6^1 = 6$

$6^2 = 6 \times 6 = 36$

$6^3 = 6 \times 6 \times 6 = 216$

$6^4 = 6 \times 6 \times 6 \times 6 = 1{,}296$

$\left(\frac{1}{3}\right)^1 = \frac{1}{3}$

$\left(\frac{1}{3}\right)^2 = \frac{1}{3} \times \frac{1}{3} = \frac{1}{9}$

$\left(\frac{1}{3}\right)^3 = \frac{1}{3} \times \frac{1}{3} \times \frac{1}{3} = \frac{1}{27}$

$\left(\frac{1}{3}\right)^4 = \frac{1}{3} \times \frac{1}{3} \times \frac{1}{3} \times \frac{1}{3} = \frac{1}{81}$

The powers of 6 are getting larger. They diverge.

The powers of $\frac{1}{3}$ are getting smaller. They converge.

Complete to find the first three powers of the given number. Tell whether the powers converge or diverge.

1. $11^1 =$ __11__

   $11^2 =$ __11__ $\times$ __11__ $=$ __121__

   $11^3 =$ __11__ $\times$ __11__ $\times$ __11__ $=$ __1,331__

   The powers of 11 __diverge__.

2. $\left(\frac{1}{7}\right)^1 =$ __$\frac{1}{7}$__

   $\left(\frac{1}{7}\right)^2 = \frac{1}{7} \times \frac{1}{7} =$ __$\frac{1}{49}$__

   $\left(\frac{1}{7}\right)^3 =$ __$\frac{1}{7}$__ $\times$ __$\frac{1}{7}$__ $\times$ __$\frac{1}{7}$__ $=$ __$\frac{1}{343}$__

   The powers of $\frac{1}{7}$ __converge__.

3. $4^1 =$ __4__

   $4^2 =$ __4__ $\times$ __4__ $=$ __16__

   $4^3 =$ __4__ $\times$ __4__ $\times$ __4__ $=$ __64__

   The powers of 4 __diverge__.

4. $\left(\frac{2}{3}\right)^1 =$ __$\frac{2}{3}$__

   $\left(\frac{2}{3}\right)^2 = \frac{2}{3} \times \frac{2}{3} =$ __$\frac{4}{9}$__

   $\left(\frac{2}{3}\right)^3 =$ __$\frac{2}{3}$__ $\times$ __$\frac{2}{3}$__ $\times$ __$\frac{2}{3}$__ $=$ __$\frac{8}{27}$__

   The powers of $\frac{2}{3}$ __converge__.

TAKE ANOTHER LOOK

LESSON 19.1

Exploring Patterns in Decimals

To change a fraction to a decimal, divide the numerator by the denominator.

$$\frac{1}{8} = 8\overline{)1.000}$$ giving 0.125, Remainder is 0.

$$\frac{5}{6} = 6\overline{)5.000}$$ giving $0.833\ldots$, or $0.8\overline{3}$. Pattern is that 3 repeats. Remainder will never be 0.

- When you divide and the remainder is zero, the decimal is a *terminating decimal*.
- When you divide and the remainder is not zero, the decimal is a *repeating decimal*.

Match the fraction with the equivalent decimal.

1. $\frac{5}{3}$ __b__   a. $2.\overline{3}$

2. $\frac{4}{12}$ __f__   b. $1.\overline{6}$

3. $\frac{8}{5}$ __e__   c. $0.\overline{4}$

4. $\frac{6}{11}$ __d__   d. $0.\overline{54}$

5. $\frac{4}{9}$ __c__   e. 1.6

6. $\frac{7}{3}$ __a__   f. $0.\overline{3}$

7. $\frac{3}{11}$ __g__   g. $0.\overline{27}$

The fraction $\frac{1}{9}$ can be renamed as $0.111\ldots$ or $0.\overline{1}$. So, $2 \times \frac{1}{9}$, or $\frac{2}{9}$, can be renamed as $2 \times 0.111\ldots$ or $2 \times 0.\overline{1}$, which equals $0.\overline{2}$.

8. The fraction $\frac{1}{3}$ can be renamed as $0.333\ldots$ or $0.\overline{3}$.

    Use this value to find the decimal equivalent of $\frac{2}{3}$. __$0.\overline{6}$__

9. The fraction $\frac{2}{25}$ can be written as the terminating decimal 0.08.

    Use this value to find the decimal equivalent of $\frac{16}{25}$. __0.64__

TAKE ANOTHER LOOK   R79

Name _____

LESSON 19.2

Patterns in Rational Numbers

Here is a way to name a rational number between the decimals 2.12 and 2.3.
You can find the average of two numbers to find a number between them.

Step 1 Add the decimals.   **Step 2** Divide by 2 to find the average.

     2.12 + 2.3 = 4.42   4.42 ÷ 2 = 2.21

The number 2.21 is between 2.12 and 2.3.

Here is a way to name a rational number between the fractions $\frac{1}{6}$ and $\frac{2}{5}$.
If you write the two fractions as equivalent fractions with a common denominator, it will be easier to find a number between them.

Step 1
Find a common denominator.

6, 12, 18, 24, __30__, 36, ...
5, 10, 15, 20, 25, __30__, ...

Step 2
Write equivalent fractions with the common denominator.

$\frac{1}{6} \times \frac{5}{5} = \frac{5}{30}$

$\frac{2}{5} \times \frac{6}{6} = \frac{12}{30}$

Step 3
Choose any number between the two numerators.

7 is a number between 5 and 12, so $\frac{7}{30}$ is between $\frac{1}{6}$ and $\frac{2}{5}$.

Complete to find a rational number between the two decimals.

1. 1.02 and 1.035

    __1.02__ + __1.035__ = __2.055__

    __2.055__ ÷ 2 = __1.0275__

2. 0.8 and 1.22

    __0.8__ + __1.22__ = __2.02__

    __2.02__ ÷ 2 = __1.01__

Name a rational number between the two decimals. **Possible answers are given.**

3. 8.3 and 8.4 __8.35__

4. 2.54 and 2.55 __2.545__

Complete to find a rational number between the fractions. **Possible answers are given.**

5. $\frac{1}{3}$ and $\frac{3}{5}$

A common denominator is __15__.

$\frac{1}{3} = \frac{\boxed{5}}{\boxed{15}}$ and $\frac{3}{5} = \frac{\boxed{9}}{\boxed{15}}$

__$\frac{7}{15}$__ is between $\frac{1}{3}$ and $\frac{3}{5}$.

6. $\frac{1}{10}$ and $\frac{1}{7}$

A common denominator is __70__.

$\frac{1}{10} = \frac{\boxed{7}}{\boxed{70}}$ and $\frac{1}{7} = \frac{\boxed{10}}{\boxed{70}}$

__$\frac{8}{70}$__ is between $\frac{1}{10}$ and $\frac{1}{7}$.

Name a rational number between the two fractions. **Possible answers are given.**

7. $\frac{4}{9}$ and $\frac{2}{3}$ __$\frac{5}{9}$__   **8.** $\frac{1}{4}$ and $\frac{2}{5}$ __$\frac{7}{20}$__   **9.** $\frac{5}{8}$ and $\frac{6}{11}$ __$\frac{50}{88}$__   **10.** $\frac{5}{12}$ and $\frac{9}{18}$ __$\frac{17}{36}$__

R80 TAKE ANOTHER LOOK

LESSON 19.3

Name _____

Patterns in Sequences

In an arithmetic sequence, each term is found by *adding* the same positive or negative number, called the common difference, to the term before it.

$$5, \ 10, \ 15, \ 20, \ldots$$
$$+5 \ +5 \ +5$$

This sequence is *arithmetic*.
The *common difference* is 5.

You find the common difference by subtracting any term from the next term.

$(10 - 5) = (15 - 10) = (20 - 15) = 5$

In a geometric sequence, each term is found by *multiplying* the term before it by the same number, called the common ratio.

$$1, \ 3, \ 9, \ 27, \ldots$$
$$\times 3 \ \times 3 \ \times 3$$

This sequence is *geometric*.
The *common ratio* is 3.

You find the common ratio by dividing any term by the term before it.

$\frac{3}{1} = \frac{9}{3} = \frac{27}{9} = 3$

Complete the table.

	Sequence	Common Ratio, Common Difference, or None?	Geometric Sequence, Arithmetic Sequence, or Neither?
1.	3, 7, 11, 15, ... +4 +4 +4	Common difference	Arithmetic sequence
2.	1, 4, 9, 16, ... +3 +5 +7	None	Neither
3.	0.2, 0.4, 0.8, 1.6, ... ×2 ×2 ×2	Common ratio	Geometric sequence

Complete. Write the next two terms in the sequence.

4. 3, 11, 19, ... __27; 35__
   +8 +8

5. 0.5, 5, 50, ... __500; 5,000__
   ×10 ×10

6. 2.4, 12, 60, ... __300; 1,500__
   ×5 ×5

7. 24, 20, 16, ... __12; 8__
   −4 −4

TAKE ANOTHER LOOK R81

LESSON 19.4

Patterns in Exponents

Base^Exponent	= Power of 2
2^4	= 16
2^3	= 8
2^2	= 4
2^1	= 2
2^0	= 1
2^{-1}	= $\frac{1}{2}$ = $\frac{1}{2^1}$
2^{-2}	= $\frac{1}{4}$ = $\frac{1}{2^2}$
2^{-3}	= $\frac{1}{8}$ = $\frac{1}{2^3}$
2^{-4}	= $\frac{1}{16}$ = $\frac{1}{2^4}$

↓ Exponent decreases.   ↓ Power of 2 decreases.

Do you see any patterns?
- As the exponent of 2 is decreased by 1, the power of 2 is divided by 2.
- A power with a negative exponent can be rewritten as a fraction with 1 in the numerator and a *positive* exponent in the denominator.
- A negative exponent does not give a negative value to the power.

When the base is greater than 1,
- if the exponent is negative, the power is always less than 1.
- if the exponent is positive, the power is always greater than 1.

You can rewrite an expression with a negative exponent as an expression with a positive exponent. To do this, write a fraction with 1 in the numerator and a positive exponent in the denominator.

$$10^{-4} = \frac{1}{10^4} \qquad 2^{-6} = \frac{1}{2^6} \qquad 2^{-4} = \frac{1}{2^4} \qquad 8^{-3} = \frac{1}{8^3}$$

You can rewrite a fraction with 1 in the numerator as an expression with a negative exponent.

- Write the denominator in exponential form.
- Then find the reciprocal and change the sign of the exponent from positive to negative.

$$\frac{1}{10{,}000} = \frac{1}{10^4} = 10^{-4} \qquad \frac{1}{625} = \frac{1}{5^4} = 5^{-4} \qquad \frac{1}{729} = \frac{1}{9^3} = 9^{-3}$$

Write each expression, using a positive exponent.

1. 7^{-2} = $\frac{1}{7^2}$
2. 10^{-9} = $\frac{1}{10^9}$
3. 8^{-4} = $\frac{1}{8^4}$
4. 3^{-6} = $\frac{1}{3^6}$
5. 4^{-10} = $\frac{1}{4^{10}}$
6. 9^{-2} = $\frac{1}{9^2}$

Complete to write each fraction, using a negative exponent.

7. $\frac{1}{64} = \frac{1}{\boxed{8^2}} =$ __$8^{-2}, 4^{-3}$, or 2^{-6}__

8. $\frac{1}{6 \times 6 \times 6} = \frac{1}{\boxed{6^3}} =$ __6^{-3}__

9. $\frac{1}{4^3} =$ __4^{-3}__

10. $\frac{1}{32} = \frac{1}{\boxed{2^5}} =$ __2^{-5}__

R82 **TAKE ANOTHER LOOK**

LESSON 20.1

Choosing a Sample

When you take a poll or survey, you can't survey everyone, so you survey a *sample*, or part of a group.

If you select wisely, the sample can be used to represent the entire *population*, or everyone.

- Everyone has an *equal chance* of being selected for a *random sample*.

  For example, you could choose any 50 seventh graders from a list of all seventh graders in your town.

- For a *systematic sample*, you randomly select the first person or object. Then you use a *pattern* to select the rest of the sample.

  For example, you could choose every fifth student from a list of all students in your town.

- For a *stratified sample*, you *divide* the population into smaller groups that contain similar people or objects. Then you randomly select a sample from each smaller group.

  For example, you could divide all students in your town into those who ride a bus, those who walk, and those who ride a bicycle to school. Then randomly select a few students from each group.

Random Sample	Systematic Sample	Stratified Sample
Equal chance	Pattern	Divided population

Determine the type of sample.

1. Twenty adults and twenty teens are asked to name their favorite author.

   _____**stratified sample**_____

2. Every fifth person who leaves a hospital is asked about health care.

   _____**systematic sample**_____

3. From a list of 300 teachers who attend an educational conference, 50 teachers are chosen.

   _____**random sample**_____

4. People who live in odd-numbered apartments are asked to name their favorite radio program.

   _____**systematic sample**_____

5. Every tenth driver who stops at a toll booth is asked about the type of gasoline the car uses.

   _____**systematic sample**_____

6. Ten women and ten men are asked to name their favorite vacation spot.

   _____**stratified sample**_____

7. Give an example of a random sample. **Answers will vary.**

_____

_____

TAKE ANOTHER LOOK   R83

Name _____

LESSON 20.2

Bias in Samples

At a county fair, 150 people show dogs and 90 people show cats.
- In a survey about pet food, to be fair or *unbiased*, the sample must be chosen proportionally from both dog and cat owners.
- The survey would be *biased* if *only* cat or *only* dog owners were sampled.

A grocery store wants to know if customers clip coupons. Which of the following sampling approaches is *not* biased?

   a. randomly surveying 50 teenagers in the store
   b. randomly surveying 50 males who buy a product on sale
   c. randomly surveying 50 shoppers at the checkout register

- Choice a is biased because it surveys only teenagers.
- Choice b is biased because it surveys only males.
- Choice c is not biased because the sample could equally represent all members of the shopping population.

Circle the sampling method that is *not* biased. Then complete to explain why the other two methods are biased.

1. A tire manufacturer wants to know how much money car owners are willing to pay for a new tire.
   (a.) randomly surveying 50 drivers who renew their car registrations
   b. randomly surveying 50 women whose license numbers contain a 7
   c. randomly surveying 50 car owners over the age of 35

   Choice __b__ is biased because it does not include __men__. Choice __c__ is biased because it does not include __car owners 35 or under__.

2. A candidate for mayor wants to know what voters think about taxes.
   a. randomly surveying every tenth voter on a college campus
   (b.) randomly surveying 50 people over the age of 18
   c. randomly surveying 100 homeowners

   Choice __a__ is biased because it does not include voters who do not __attend college__. Choice __c__ is biased because it does not include voters who do not __own their own homes__.

R84 TAKE ANOTHER LOOK

Name _____

LESSON 20.3

Writing Survey Questions

The purpose of a survey is to collect information from a sample of a population. This information can help you make predictions.

For example, you can use survey results to predict the buying patterns or voting patterns of a population.

There are many different ways to ask questions on a survey.

- A *short-answer question* is open-ended. You can answer some short-answer questions with "yes" or "no." Others need a word or two.

    Examples: "Do you enjoy listening to country music?"

    "What is your favorite style of music?"

- A *multiple-choice question* gives you answers from which to choose. You must select one of the answer choices.

    Example: "Which style of music do you prefer?"

    a. rock   **b.** country   **c.** classical   **d.** jazz

- A *fill-in-the-blank question* is a sentence you complete with a word or words.

    Example: "When you listen to music, you prefer to listen to _____."

- A *numerical question* asks you to rate an item on a numerical scale. The question should tell you which number is high and which is low.

    Example: "On a scale of 1 to 10, with 1 meaning never and 10 meaning always, how often do you listen to country music?"

Name the format of the survey question.

1. What is your favorite job at home?

    _____ short answer _____

2. Your favorite television program

    is _____ fill-in-the-blank _____.

3. On a scale of 1 to 5, with 1 being seldom and 5 being often, how many times do you go to the movies every month?

    _____ numerical _____

4. Which drink do you prefer?

    a. water   **b.** milk   **c.** juice   **d.** soda

    _____ multiple choice _____

5. Your favorite professional baseball team

    is _____ fill-in-the-blank _____.

6. What is your favorite color?

    _____ short answer _____

7. What is your favorite color?

    a. red   **b.** green   **c.** blue   **d.** yellow

    _____ multiple choice _____

8. On a scale of 1 to 5, with 1 being needs improvement and 5 being wonderful, how would you rate your school's baseball team?

    _____ numerical _____

TAKE ANOTHER LOOK

Name _____

LESSON 20.4

Organizing and Displaying Results

There are many different ways to display data. A stem-and-leaf plot is one way. The stems in a stem-and-leaf plot are the tens or tens and hundreds digits. The leaves are the corresponding ones digits.

Make a stem-and-leaf plot of the data in the table.

Circumference of Pumpkins (in cm)									
156	147	161	139	166	160	155	158	156	137

Step 1 Draw a vertical line. Label the left side *Stem* and the right side *Leaves*.

Stem	Leaves

156, 147, 161, 139, 166, 160, 155, 158, 156, 137

Step 2 Underline the tens and hundreds digits in the data. Order them from least to greatest. Write them in the *Stem* column. Include the digits only once if they occur more than once.

Stem	Leaves
13	
14	
15	
16	

Step 3 Write the corresponding ones digits in order from least to greatest in the *Leaves* column. If a digit appears more than once, write it as many times as it appears. Make a key.

Stem	Leaves
13	7 9
14	7
15	5 6 6 8
16	0 1 6

13|7 represents 137

Step 4 Compare the number of data in the plot with the number of data in the table to make sure you have included all the data.

Make a stem-and-leaf plot of the data in the table.

Weight of Rock Samples (in g)						
98	86	76	80	100	82	92
100	81	98	85	77	93	89
76	98	75	100	84	79	92

Stem	Leaves
7	5 6 6 7 9
8	0 1 2 4 5 6 9
9	2 2 3 8 8 8
10	0 0 0

7|5 represents 75

R86   TAKE ANOTHER LOOK

Name _____

LESSON 21.1

How Do Your Data Shape Up?

A stem-and-leaf plot can help you organize data and make a histogram.
Use a stem-and-leaf plot to organize the following data. Then make a histogram.

math test scores: 59, 83, 86, 72, 61, 94, 70, 68, 53, 98, 81, 70, 68, 81, 96, 73, 80, 70

Stem	Leaves		Interval		Number of Scores
5	3 9	→	50 to 59	→	2
6	1 8 8	→	60 to 69	→	3
7	0 0 0 2 3	→	70 to 79	→	5
8	0 1 1 3 6	→	80 to 89	→	5
9	4 6 8	→	90 to 99	→	3

Stem-and-Leaf Plot

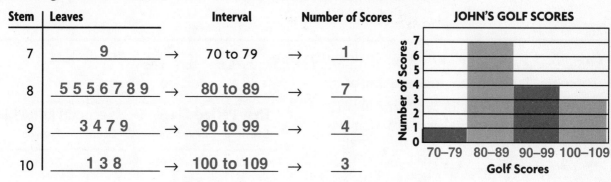

Histogram — MATH TEST SCORES

---

1. Complete to construct a stem-and-leaf plot and a histogram for the data.

   John's golf scores: 93, 85, 85, 103, 97, 87, 88, 86, 94, 99, 101, 85, 89, 108, 79

Stem	Leaves		Interval		Number of Scores
7	9	→	70 to 79	→	1
8	5 5 5 6 7 8 9	→	80 to 89	→	7
9	3 4 7 9	→	90 to 99	→	4
10	1 3 8	→	100 to 109	→	3

JOHN'S GOLF SCORES histogram

For Exercises 2–8, use the stem-and-leaf plot and histogram above.

2. What was John's highest score? __108__

3. What was John's lowest score? __79__

4. What is the range of scores? __29__

5. Which interval has the greatest number of scores? __80 to 89__

6. Which interval has the least number of scores? __70 to 79__

7. How many times did John score at least 90? __7 times__

8. Which graph gives more information about John's actual scores? Explain.

   __Stem-and-leaf plot; it shows actual scores.__

TAKE ANOTHER LOOK R87

LESSON 21.2

Central Tendencies

Mrs. Smith gave her science class a test. Now she wants to determine how the class as a whole did on the test.

science test scores: 90, 52, 70, 73, 82, 93, 67, 78, 87, 92, 71, 43, 79, 75, 87

- She takes the scores and finds the measures of **central tendency**. Then she must decide which measure best represents the data.
- First, she organizes the data in a stem-and-leaf plot.
- The *mean* is the sum of the scores divided by the number of scores.

    mean: $\frac{1{,}139}{15} \approx 76$
- The *mode* is the score(s) that occurs most frequently.

    mode: 87
- The *median* is the middle number when the data are arranged in order. If there is an even number of scores, average the two middle numbers.

    median: 78

Stem	Leaves
4	3
5	2
6	7
7	0 1 3 5 8 9
8	2 7 7
9	0 2 3

The mean and the median are close. So, both are good measures of central tendency to represent the data.

Mr. Johnson gave his first- and second-period classes the same test. The scores are shown in the stem-and-leaf plots.

1. What is the mean for Mr. Johnson's first-period class? for his second-period class?

    _____72; 78_____

2. What is the median for Mr. Johnson's first-period class? for his second-period class?

    _____70; 81_____

3. What is the mode for Mr. Johnson's first-period class? for his second-period class?

    _____100; 78 and 85_____

First-Period Class

Stem	Leaves
4	0
5	3 5 8
6	0 2 5 6 9
7	0 2 5 5
8	1 2 5
9	
10	0 0 0

Second-Period Class

Stem	Leaves
3	2 9
4	
5	
6	8
7	0 5 8 8 8 9
8	0 2 4 5 5 5 6
9	0 3 5 8

4. The measures of central tendency help Mr. Johnson compare classes. What is a good measure of central tendency? Explain.

    _____Mean or median; both are central values._____

5. Which class scored better? Explain.

    _____The second-period class; the mean and the median were both higher._____

Name _____

LESSON 21.3

Using Appropriate Graphs

The table shows how teenagers spend their allowances. What type of graph would best display this data?

Clothing	Food	Entertainment	Other
27%	37%	24%	12%

- The data do not show a change over time, so a *line graph* would not be best.
- You are not comparing data sets, so *multiple box-and-whisker graphs* would not be best.
- Since the data are grouped in categories, you could use a *bar graph*. However, the bar graph does not show how all categories are related to one another.

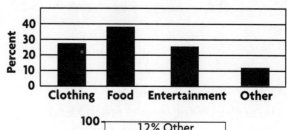

- Since the data are given as percents whose sum equals 100%, a *stacked bar graph* best displays the data.

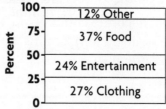

The table below shows the population of Merryville between 1930 and 1990.

Year	1930	1940	1950	1960	1970	1980	1990
Population (in thousands)	137	561	800	690	640	565	580

1. Are data given as percents? __no__
2. Are data categorical? __no__
3. Do data show changes over time? __yes__
4. Does the table compare sets of data? __no__
5. Which type of graph would be best to represent the data? Why?

   __Line graph; data show changes over time.__

The table shows the hours per day seventh graders at Marshall School spend on homework.

Hours	0	<1	1–2	>2
Students	35	38	16	10

6. Are data given as percents? __no__
7. Are data categorical? __yes__
8. Do data show changes over time? __no__
9. Does the table compare sets of data? __no__
10. Which type of graph would be best? Why?

    __Bar graph; data are categorical.__

TAKE ANOTHER LOOK R89

Name _____

LESSON 21.4

Misleading Graphs

The band and orchestra at Jenkins Junior High want to see who sold more candy bars over a 4-week period.

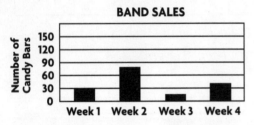
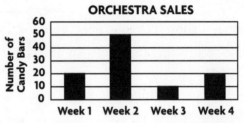

- The graphs make it appear that the orchestra sold more candy bars than the band.
- When you read the scales, you see this is not true.
- The scales on the graphs are misleading.
- To make an accurate comparison, either make the scales the same on both graphs, or make a double-bar graph.

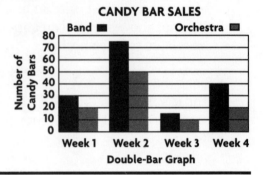

For Exercises 1–5, use the graphs at the right.

1. Which team scored the most points? __**Cardinals**__

2. Which team scored the fewest points? __**Sparrows**__

3. What is the ratio of points scored by Bluejays to points scored by Sparrows? What ratio do the bars on the graphs reflect?

   __**6 to 1; about 2 to 1**__

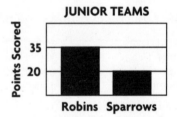

4. What is misleading about these graphs? __**the scales**__

5. How would you graph an accurate comparison of the points scored by the Junior and Senior Teams?

   __**Possible answer: Make one double-bar graph,**__

   __**or make the scales the same.**__

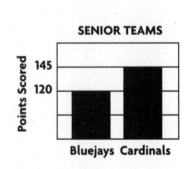

For Problems 6–7, use the line graph at the right.

6. What is the approximate ratio of the dollar values of computers sold to televisions sold in 1995? But what ratio does the graph show?

   __**about 4 to 1; about 2 to 1**__

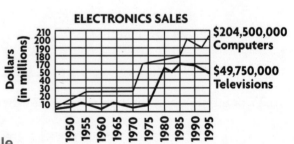

7. What is misleading about this graph? __**the scale**__

R90 TAKE ANOTHER LOOK

Name _____

LESSON 22.1

Tree Diagrams and Sample Spaces

Doug's work clothes consist of 2 pairs of pants and 4 shirts. When Doug gets dressed, he can choose either black or brown pants. He can then choose either a blue, tan, green, or orange shirt. How many different outfits can Doug wear?

Make a tree diagram of the sample space.

Step 1 What color shirts can Doug wear with black pants?

Pants	Shirt	Possible Outcome
black	blue	black, blue
	tan	black, tan
	green	black, green
	orange	black, orange

Step 2 What color shirts can Doug wear with brown pants?

Pants	Shirt	Possible Outcome
brown	blue	brown, blue
	tan	brown, tan
	green	brown, green
	orange	brown, orange

Doug can wear 4 different shirts with black pants and 4 different shirts with brown pants. There are 8 possible outfits he can wear.

choice of pants	×	choice of shirts	=	total possible choices
2 ways		4 ways		8 ways

Complete to find the total number of possible outcomes.

The school cafeteria offers turkey or ham sandwiches. You can choose a cookie, pudding, or fruit for dessert. How many lunch choices are there, if you have a sandwich and a dessert?

Sandwich	Dessert	Possible Outcome
turkey	cookie	turkey, cookie
	pudding	turkey, pudding
	fruit	turkey, fruit
ham	cookie	ham, cookie
	pudding	ham, pudding
	fruit	ham, fruit

choice of sandwich	×	choice of dessert	=	total possible choices
2		3		6

TAKE ANOTHER LOOK R91

Name _____

LESSON 22.2

Finding Probability

You put 4 blue, 5 red, 3 yellow, and 3 green marbles in a bag. You reach in the bag without looking and pick one marble.

What is the probability of picking a blue marble?

You can use a ratio to find a probability.

$$\text{Probability (P)} = \frac{\text{number of favorable outcomes}}{\text{number of possible outcomes}}$$

$$\text{P(blue)} = \frac{4}{15} \begin{array}{l} \leftarrow \text{number of blue marbles} \\ \leftarrow \text{total number of marbles} \end{array}$$

The probability of picking a blue marble is $\frac{4}{15}$.

What is the probability of picking a yellow *or* a green marble?

$$\text{P(yellow or green)} = \frac{3+3}{15} \begin{array}{l} \leftarrow \text{number of yellow } or \text{ green marbles} \\ \leftarrow \text{total number of marbles} \end{array}$$

The probability of picking a yellow or a green marble is $\frac{6}{15}$, or $\frac{2}{5}$.

What is the probability of picking an orange marble?

$$\text{P(orange)} = \frac{0}{15} \begin{array}{l} \leftarrow \text{number of orange marbles} \\ \leftarrow \text{total number of marbles} \end{array}$$

The probability of picking an orange marble is 0.

All sections of the spinner are equal. Complete to find the probability.

1. $\text{P(5)} = \dfrac{\boxed{1}}{\boxed{8}} \begin{array}{l} \leftarrow \text{sections labeled 5} \\ \leftarrow \text{total number of sections} \end{array}$ $\text{P(5)} = \dfrac{1}{8}$

2. $\text{P(odd number)} = \dfrac{\boxed{4}}{\boxed{8}} \begin{array}{l} \leftarrow \text{number of odd-numbered sections} \\ \leftarrow \text{total number of sections} \end{array}$ $\text{P(odd number)} = \dfrac{4}{8}, \text{ or } \dfrac{1}{2}$

A bag of jelly beans contains 6 black, 4 red, 5 pink, 3 white, and 8 yellow jelly beans. One jelly bean is chosen. Complete to find the probability.

3. $\text{P(black)} = \dfrac{\boxed{6}}{\boxed{26}} \begin{array}{l} \leftarrow \text{number of black jelly beans} \\ \leftarrow \text{total number of jelly beans} \end{array}$ $\text{P(black)} = \dfrac{6}{26}, \text{ or } \dfrac{3}{13}$

4. $\text{P(red or yellow)} = \dfrac{\boxed{12}}{\boxed{26}} \begin{array}{l} \leftarrow \text{number of red } or \text{ yellow jelly beans} \\ \leftarrow \text{total number of jelly beans} \end{array}$ $\text{P(red or yellow)} = \dfrac{12}{26}, \text{ or } \dfrac{6}{13}$

R92 TAKE ANOTHER LOOK

Name _____

LESSON 22.3

Problem-Solving Strategy

Make a List: Combinations and Probability

Max is a freshman at Fairview Community College. The college offers 5 classes to freshmen: Algebra, English, Chemistry, Humanities, and Spanish. Max can take 3 classes. How many combinations of 3 classes can he take?

- How many classes can Max take? 3 classes
- How many classes can Max choose from? 5 classes
- Make a list of all possible combinations of 3 classes. Remember, *Algebra, English, Chemistry* is the same as *Chemistry, English, Algebra.*

Combinations beginning with Algebra	Combinations beginning with English that are not in the first column	Combinations that are not in the first and second columns
Algebra, English, Chemistry	English, Chemistry, Humanities	Chemistry, Humanities, Spanish
Algebra, English, Humanities	English, Chemistry, Spanish	
Algebra, English, Spanish	English, Humanities, Spanish	
Algebra, Chemistry, Humanities		
Algebra, Chemistry, Spanish		
Algebra, Humanities, Spanish		
6 combinations	3 combinations	1 combination

There are 6 + 3 + 1, or 10, different combinations of 3 classes that Max can take.

Complete to make a list and solve.

1. Alice, Bob, Chuck, Dave, and Edith have volunteered to serve in the student government. How many ways can 2 of them be selected for the Yearbook Committee?

Beginning with

Alice	**Bob**	**Chuck**	**Edith**
Alice and Bob	Bob and Chuck	__Chuck__ and __Edith__	
Alice and Chuck	__Bob__ and __Edith__		
__Alice__ and __Edith__			

There are __3__ + __2__ + __1__ + __0__ = __6__ different ways to select two students.

TAKE ANOTHER LOOK R93

Name _____

LESSON 22.4

Finding Permutations and Probability

Fix-Your-Car Co. organizes a car derby. Conquerer, Twice Hit, Lucky Lady, No Frills, and Speedy are the entries. If there are no ties, in how many ways can the 5 cars come in first, second, and third places?

 5 choices    4 choices    3 choices left

$5 \times 4 \times 3 = 60$ ways

Complete to solve.

1. In how many ways can 7 people line up at a checkout counter of a supermarket?

7	6	5	4	3	2	1
first in line	second	third	fourth	fifth	sixth	seventh

   $7 \times 6 \times 5 \times 4 \times \underline{3} \times \underline{2} \times \underline{1} = \underline{5{,}040}$

   There are __5,040__ ways for 7 people to line up at a checkout counter.

2. Manny is taking 5 classes. Each has one textbook. In how many ways can he stack the 5 books he carries to school?

5	4	3	2	1
bottom book	next	next	next	top book

   $\underline{5} \times \underline{4} \times \underline{3} \times \underline{2} \times \underline{1} = \underline{120}$

   Manny can stack the 5 books in __120__ ways.

3. How many 3-digit numbers can Jolene form from the digits 1, 3, 5, 7, and 9, with no digit repeated?

5	4	3
first digit	second digit	third digit

   $\underline{5} \times \underline{4} \times \underline{3} = \underline{60}$

   Jolene can form __60__ three-digit numbers.

4. Mario, Bill, Kevin, Nate, Greg, and Tom enter a marathon. If there are no ties, in how many ways can the first 2 prizes be awarded?

6	5
first place	second place

   $\underline{6} \times \underline{5} = \underline{30}$

   There are __30__ ways the first 2 prizes can be awarded.

5. Hoosier Falls Middle School has 12 doors. In how many ways can a student enter by one door and exit by a different door?

   132 ways

R94 TAKE ANOTHER LOOK

Experimental Probability

LESSON 23.1

Ricki has a spinner with 4 spaces of equal size colored red, blue, green, and yellow. She knows that the mathematical probability of the pointer stopping in the red space is $\frac{1}{4}$.

Ricki wants to compare the mathematical probability with the experimental probability. She spins the pointer 50 times and records her results in a table.

Color	Red	Blue	Green	Yellow
Times spun	14	18	7	11

Ricki finds the *experimental probability* of landing on red.

experimental probability = $\frac{\text{number of times event occurs}}{\text{total number of trials}}$

experimental probability of landing on red = $\frac{14 \text{ red spins}}{50 \text{ total spins}}$, or $\frac{7}{25}$

Complete to find the experimental probability of stopping on the other sections.

1. experimental probability of landing on blue = $\frac{18 \text{ blue spins}}{50 \text{ total spins}}$, or $\frac{9}{25}$

2. experimental probability of landing on green = $\frac{7 \text{ green spins}}{50 \text{ total spins}}$

3. experimental probability of landing on yellow = $\frac{11 \text{ yellow spins}}{50 \text{ total spins}}$

Now Ricki rolls a number cube labeled 1–6. She rolls the cube 60 times and records her results in a table.

Number	1	2	3	4	5	6
Times rolled	8	15	6	12	9	10

Find Ricki's experimental probability of rolling the number on the number cube.

4. number 1 $\frac{2}{15}$ 5. number 2 $\frac{1}{4}$ 6. number 3 $\frac{1}{10}$

7. number 4 $\frac{1}{5}$ 8. number 5 $\frac{3}{20}$ 9. number 6 $\frac{1}{6}$

TAKE ANOTHER LOOK

Problem-Solving Strategy

Acting It Out by Using Random Numbers

Sherman's General Store is holding a contest.

Each customer who spends over $35 is given a token with a number 1–5. Whoever collects one of each type of token wins $25 in store credit.

Gina wants to know the number of purchases over $35 she has to make to get the five different tokens and win the store credit. She uses a table of 50 random numbers to act out a probability experiment.

The table shows 50 random numbers 1–5.

1	2	2	3	4	4	3	4	4	3
4	4	4	4	1	1	2	5	5	4
2	4	2	3	2	1	2	2	2	2
5	2	3	3	1	3	3	3	2	5
4	2	4	3	1	3	1	1	1	3

Step 1

- Each number in the table represents 1 purchase. Gina counts how many numbers it will take to get all the numbers 1–5.
- Since the numbers in the table are random, Gina can begin counting anywhere. She begins with the third line.
- Gina counts 11 numbers before getting numbers 1–5. She must make 11 purchases to get all 5 tokens.

Step 2

- Gina chooses another starting point in the table. She begins at the last row and returns to the beginning of the table.
- She counts 28 numbers before getting numbers 1–5.
- Counting this way, Gina must make 28 purchases to get all 5 tokens.

Step 3

- Gina averages the results to get a better estimation. $\frac{11 + 28}{2} = 19.5$

So, Gina needs to make about 20 purchases over $35 to win $25 in store credit.

Use the table of random numbers. Choose two different starting points for each exercise. List the starting points you select. **Answers will vary.**

1. A service station gives away a free glass with each purchase of 8 gal or more of gasoline. There are 5 different glass designs. How many purchases will you have to make to get one of each design?

2. A fast food restaurant includes a free cartoon figure with each child's meal. There are 5 different cartoon figures. How many meals will you need to buy to get one of each?

Name _____

LESSON 23.3

Designing a Simulation

Scientists would like to count the number of polar bears on the Seward Peninsula. Because they cannot count every bear, they design a *simulation*, or a model of an investigation that would be too difficult or time-consuming to perform.

One month, the scientists trap, tag, and release 200 polar bears. The next month, they trap and examine 150 polar bears.

Because 16 polar bears have tags, scientists can *estimate* the total number of polar bears on the Seward Peninsula.

Step 1 Scientists use a proportion.

$$\frac{\text{tagged polar bears trapped during 2nd month}}{\text{total trapped polar bears during 2nd month}} = \frac{\text{total polar bears tagged}}{\text{total polar bears on Seward Peninsula}}$$

Let x represent the total number of polar bears on Seward Peninsula.

Step 2 Set up and solve the proportion. $\qquad \frac{16}{150} = \frac{200}{x}$

Cross multiply. $\qquad \frac{16}{150} \diagup\!\!\!\!\diagdown \frac{200}{x}$

Solve for x. $\qquad 16x = 30{,}000,\ \text{or}\ x = 1{,}875$

A simulation can only estimate the actual polar bear count. There are *about* 1,875 polar bears on Seward Peninsula.

Complete to find the number of fish in each lake.

At Lake Wright, 150 fish are caught, tagged, and released. The *same number*, 150 fish, are caught and released the next day, and 30 fish have tags.

1. What proportion do you use to solve the problem?

$$\frac{\boxed{\text{tagged fish}}\ \text{caught on 2}^{\text{nd}}\ \text{day}}{\text{total}\ \boxed{\text{tagged fish}}} = \frac{\boxed{\text{total fish}}\ \text{caught on 2}^{\text{nd}}\ \text{day}}{\text{total}\ \boxed{\text{fish in lake}}}$$

2. Set up and solve the proportion to find about how many fish are in Lake Wright.

_____ **about 750 fish** _____

At Harmony Lake, 175 bass are caught, tagged, and released. The *same number*, 175 bass, are caught and released the next day, and 25 have tags.

3. What proportion do you use to solve the problem?

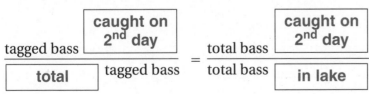

4. About how many bass are in Harmony Lake? _____ **about 1,225 bass** _____

TAKE ANOTHER LOOK

Name _____

LESSON 23.4

Geometric Probability

The surface of a swimming pool must be skimmed so leaves will not clog the filter. What is the *geometric probability* that a leaf blowing onto a 100-ft × 100-ft yard will randomly land in a circular pool with a radius of 15 ft? Use 3.14 for π and round to hundredths.

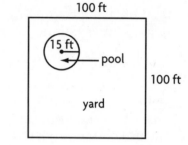

Step 1 Find the "landing" area.
area of pool = $\pi(15)^2 \approx 3.14 \times 15 \times 15 \approx 706.5$ ft^2

Step 2 Find the area of the yard.
area of yard = $100 \times 100 = 10,000$ ft^2

Step 3 Set up a ratio.

$\dfrac{\text{area of favorable outcome}}{\text{total area}} \rightarrow \dfrac{\text{area of pool}}{\text{area of yard}} \approx \dfrac{706.5}{10,000} \approx 0.07$

The geometric probability that a leaf blowing onto the yard will land in the pool is about $\dfrac{7}{100}$, 0.07, or 7%.

Complete to find the geometric probability that a leaf blowing onto a 75-ft × 50-ft yard will land in a circular fish pond with a radius of 5 ft.

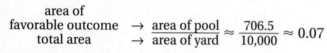

1. Find the area of the fish pond. ____**78.5 ft²**____

2. Find the area of the yard. ____**3,750 ft²**____

3. geometric probability =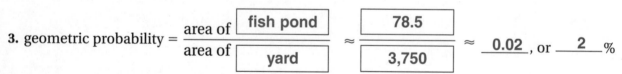

Complete to find the geometric probability that a leaf blowing onto a circular yard with a radius of 60 ft will land in a 30-ft × 20-ft rectangular pool.

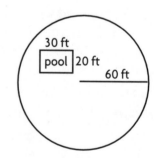

4. Find the area of the pool. ____**600 ft²**____

5. Find the area of the yard. ____**11,304 ft²**____

6. geometric probability =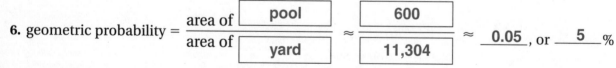

TAKE ANOTHER LOOK

Name _____

LESSON 24.1

Measuring and Estimating Lengths

Suppose you are a heart surgeon measuring an artery. Would you measure in millimeters or in centimeters?

- Ask yourself: How *precise* must the measurement be?
- An artery measurement should be as precise as possible.
- From smaller to larger, metric units of measurement are:
  mm   cm   dm   m   dam   hm   km
- Smaller units give a more precise measurement.

You should measure the artery in millimeters because they are smaller units than centimeters.

- An object measured to the nearest inch has a *precision* of 1 in.
- An object measured to the nearest $\frac{1}{4}$ in. has a precision of $\frac{1}{4}$ in.
- The precision of a $2\frac{3}{4}$-mm artery is $\frac{1}{4}$ mm.

Which measurement is more precise? Circle your answer.

1. 15 km or (14,500 m)
2. $3\frac{1}{2}$ yd or (10 ft)
3. (5,000 ft) or 1 mi
4. 2 ft or (27 in.)
5. (17 in.) or $1\frac{1}{2}$ ft
6. 12 cm or (115 mm)

Complete to give the precision of each measurement.

	Item	Precision
7.	8-oz steak	__1__ oz
8.	$4\frac{3}{4}$-in. bolt	$\frac{1}{4}$ in.
9.	$4\frac{3}{8}$-yd piece of fabric	$\frac{1}{8}$ yd
10.	26-mi race track	__1__ mi
11.	8-m high wall	__1__ m
12.	3,505 nautical mi	__1__ nautical mi
13.	7.5 light years	__0.1__ light year
14.	2.58-cm scar	__0.01__ cm

TAKE ANOTHER LOOK

Name _____

LESSON 24.2

Networks

Three families are moving from Lubbock. The Dinardos are moving to Dallas, the LaCavas to Austin, and the Amicks to San Antonio.

Ace Moving and Storage packs all their belongings into one moving van.

There are six possible routes from Lubbock through San Antonio, Austin, and Dallas. Mr. Sams, the driver, wants to save time and money. Use a network to help him find the shortest route.

Complete the missing routes and distances.

	Route	Distance
1.	Lubbock, Dallas, __Austin__, San Antonio	340 + 196 + 80 = 616 mi
2.	Lubbock, Austin, __San Antonio__, Dallas	450 + 80 + 270 = 800 mi
3.	Lubbock, San Antonio, __Dallas__, Austin	400 + 270 + 196 = 866 mi
4.	Lubbock, __San Antonio__, __Austin__, Dallas	400 + 80 + 196 = 676 mi
5.	Lubbock, __Austin__, __Dallas__, San Antonio	450 + 196 + 270 = 916 mi
6.	Lubbock, __Dallas__, __San Antonio__, Austin	340 + 270 + 80 = 690 mi

7. Which route is the shortest? What is the distance?

   __Lubbock, Dallas, Austin, San Antonio; 616 mi__

8. Complete the routes and distances from Charlotte through each of the other three cities. Circle the longest route.

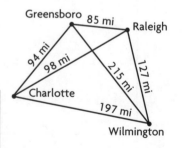

Route	Total Distance
Charlotte, Greensboro, Raleigh, Wilmington	306 mi
Charlotte, Raleigh, Wilmington, Greensboro	440 mi
(Charlotte, Wilmington, Greensboro, Raleigh)	497 mi
Charlotte, Wilmington, Raleigh, Greensboro	409 mi
Charlotte, Raleigh, Greensboro, Wilmington	398 mi
Charlotte, Greensboro, Wilmington, Raleigh	436 mi

R100 TAKE ANOTHER LOOK

Name _____

LESSON 24.3

Pythagorean Property

In a right triangle, the side opposite the right angle is the *hypotenuse* and the other two sides are *legs*. The hypotenuse is always the longest side.

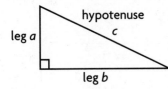

The Pythagorean Property is the relationship
$(\text{leg})^2 + (\text{leg})^2 = (\text{hypotenuse})^2$.
$\quad a^2 \;\;+\;\; b^2 \;\;=\;\; c^2$

Use the formula to find the hypotenuse of the triangle at the left.

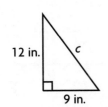

$12^2 + 9^2 = c^2$
$144 + 81 = c^2$
$225 = c^2$
$\sqrt{225} = \sqrt{c^2}$ ← Find the square root of both sides.
$15 = c \qquad$ The hypotenuse is 15 in.

Complete to find the length of the hypotenuse to the nearest tenth. Use a calculator to help you find the square roots.

1.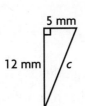

$c^2 = 5^2 + 12^2$

$c^2 = \underline{\;\;25\;\;} + \underline{\;\;144\;\;}$

$c^2 = \underline{\;\;169\;\;}$

$c = \sqrt{\underline{\;\;169\;\;}}$

$c = \underline{\;\;13\;\;}$

The hypotenuse is __13__ mm.

2.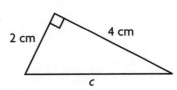

$c^2 = \boxed{2}^2 + \boxed{4}^2$

$c^2 = \underline{\;\;4\;\;} + \underline{\;\;16\;\;}$

$c^2 = \underline{\;\;20\;\;}$

$c = \sqrt{\underline{\;\;20\;\;}}$

$c \approx \underline{\;\;4.5\;\;}$

The hypotenuse is approximately __4.5__ cm.

3.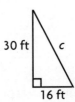

$c^2 = \boxed{16}^2 + \boxed{30}^2$

$c^2 = \underline{\;\;256\;\;} + \underline{\;\;900\;\;}$

$c^2 = \underline{\;\;1{,}156\;\;}$

$c = \sqrt{\underline{\;\;1{,}156\;\;}}$

$c = \underline{\;\;34\;\;}$

The hypotenuse is __34__ ft.

Find the length of the hypotenuse if leg *a* and leg *b* have the given measurements.

4. 24 yd, 45 yd

__51 yd__

5. 10 in., 24 in.

__26 in.__

6. 7 ft, 24 ft

__25 ft__

7. 15 cm, 20 cm

__25 cm__

TAKE ANOTHER LOOK R101

Name _____

LESSON 24.4

Problem-Solving Strategy

Using a Formula to Find the Area

What are the areas of the four figures in the figure at the right? Use formulas. Remember, area is always measured in square units.

area of a triangle = $\frac{1}{2} \times b \times h = \frac{1}{2} \times 9 \times 6 = 27$, 27 ft²
area of a parallelogram = $b \times h = 12 \times 6 = 72$, 72 ft²
area of a rectangle = $l \times w = 9 \times 8 = 72$, 72 ft²
area of a rectangle = $l \times w = 12 \times 8 = 96$, 96 ft²

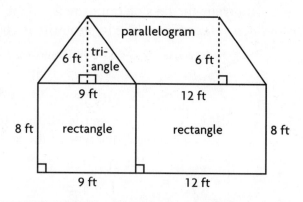

Complete to solve.

1. Marge is making a bedspread. She needs to calculate how many square feet of fabric to buy. The dimensions of the bed are 5 ft by 6 ft. How much fabric should she buy?

   Use the formula: $A =$ __lw__.

   So, $A =$ __5__ × __6__.

   She should buy __30__ ft² of fabric.

2. Mrs. Math has a triangular flower garden. What is the area of her flower garden?

   Use the formula: $A = \frac{1}{2} bh$.

   So, $A = \frac{1}{2}$ × __6__ × __8__.

   The area of the flower garden is __24__ ft².

3. Brad wants to carpet part of his basement floor. The area he wants to carpet is in the shape of a parallelogram with a height of 13 ft and a base of 16 ft. The carpet costs $8.50 per square foot. What must Brad spend to carpet his basement?

   Use the formula: $A =$ __bh__.

   So, $A =$ __13__ × __16__.

   cost = __208__ × $8.50

   The total cost for carpeting is __$1,768__.

4. Cory's father is building a sandbox for him. The bottom and four sides are rectangles. Calculate how many square feet of wood he needs for the bottom and four sides.

   area of bottom = __63 ft²__

   area of 4 sides = __7 ft²__ +

   __7 ft²__ + __9 ft²__ + __9 ft²__

   Cory's father must buy __95 ft²__ of wood.

5. Mr. Turk wants to plant grass seed to cover the area pictured. What is the total area of the lawn he wants to cover?

   __282 ft²__

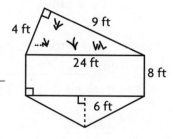

R102 **TAKE ANOTHER LOOK**

Area of a Trapezoid

The formula for the area of a triangle $A = \frac{1}{2}bh$ can help you find the area of a trapezoid.

On a separate sheet of paper, trace and cut out the trapezoid. Then cut along the dashed lines to form triangles X and Y.

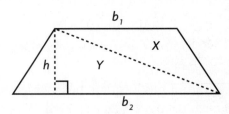

- Notice that triangles X and Y have the same height as the trapezoid.
- The areas of triangles X and Y are $\frac{1}{2}b_1 h$ and $\frac{1}{2}b_2 h$.
- So, the area of the trapezoid equals $\frac{1}{2}b_1 h + \frac{1}{2}b_2 h$.
- Using the Distributive Property, $A = \frac{1}{2}h(b_1 + b_2)$.

Find the area of the trapezoid.

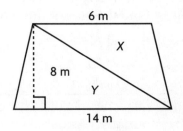

area of triangle $X = \frac{1}{2}bh = \frac{1}{2} \times 6 \times 8 = 24$

area of triangle $Y = \frac{1}{2}bh = \frac{1}{2} \times 14 \times 8 = 56$

area of trapezoid $= 24 + 56 = 80$

The area of the trapezoid is 80 m².

In Exercises 1–6, complete to find the area of the trapezoid.

1. 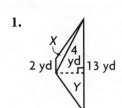   area of triangle $X =$ __4 yd²__
   area of triangle $Y =$ __26 yd²__
   area of trapezoid $=$ __30 yd²__

2.   area of triangle $X =$ __95 ft²__
   area of triangle $Y =$ __55 ft²__
   area of trapezoid $=$ __150 ft²__

3.   area of triangle $X =$ __42 in.²__
   area of triangle $Y =$ __60 in.²__
   area of trapezoid $=$ __102 in.²__

4. 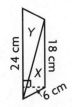   area of triangle $X =$ __54 cm²__
   area of triangle $Y =$ __72 cm²__
   area of trapezoid $=$ __126 cm²__

Write the formula for the area of a trapezoid. Then find the area.

5.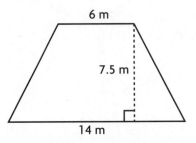

   $A = \frac{1}{2}h(b_1 + b_2)$; 75 m²

6.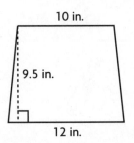

   $A = \frac{1}{2}h(b_1 + b_2)$; 104.5 in.²

Name _____

LESSON 25.1

Surface Area of Prisms and Pyramids

To find the surface area of a square pyramid, it helps to remember the following:
- A square pyramid has 1 square face and 4 triangular faces.
- To find the area of a square, use the formula $A = s^2$.
- To find the area of a triangle, use the formula $A = \frac{1}{2}bh$.
- To find the surface area of a pyramid, you must add the areas of the 5 faces.

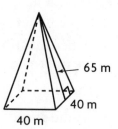

A pyramid has a square base which measures 40 m on each side. The height of each triangular face is 65 m. What is the surface area of the pyramid?

Step 1 Find the area of the square base. $A = 40 \times 40$ $= 1,600$ $= 1,600 \text{ m}^2$	Step 2 Find the area of 1 triangular face. $A = \frac{1}{2} \times 40 \times 65$ $= 1,300$ $= 1,300 \text{ m}^2$
Step 3 Find the area of 4 triangular faces. $A = 4 \times 1,300$ $= 5,200$ $= 5,200 \text{ m}^2$	Step 4 Find the sum of the areas of the 5 faces. surface area = $1,600 \text{ m}^2 + 5,200 \text{ m}^2 = 6,800 \text{ m}^2$

Find each area.

1.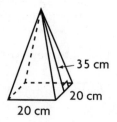

   area of square face: __400 cm²__

   area of 1 triangular face: __350 cm²__

   area of 4 triangular faces: __1,400 cm²__

   surface area: __1,800 cm²__

2.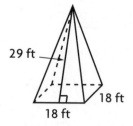

   area of square face: __324 ft²__

   area of 1 triangular face: __261 ft²__

   area of 4 triangular faces: __1,044 ft²__

   surface area: __1,368 ft²__

3.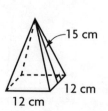

   surface area: __504 cm²__

4.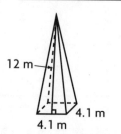

   surface area: __115.2 m²__

R104 TAKE ANOTHER LOOK

Name _____

LESSON 25.2

Finding Surface Area of Cylinders

To find the surface area of a cylinder, it helps to remember the following:
- The lateral surface of a cylinder can be unrolled to form a rectangle.
- To find the area of this rectangle, multiply the circumference of the base by the height of the cylinder.
- The formula for the circumference of a circle is $C = \pi d$.
- The formula for the area of a circle is $A = \pi r^2$.
- To find the surface area of a cylinder, add the areas of the 2 bases and the area of the lateral surface.

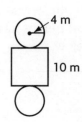

The pattern to make a cylinder is shown. Find the surface area. Use $\pi = 3.14$.

Step 1 Find the area of the lateral surface. $A = \pi d \times h$ $= 3.14 \times 8 \times 10$ $= 251.2$	**Step 2** Find the area of one base. $A = \pi r^2$ $= 3.14 \times 4^2$ $= 50.24$
Step 3 Find the area of the 2 bases. $A = 2 \times 50.24$ $= 100.48$	**Step 4** Find the sum of the areas of the 2 bases and the lateral surface. surface area $= 251.2 + 100.48$ $= 351.68$ $= 351.68 \text{ m}^2$

Complete for each cylinder. Use 3.14 for π. Round to the nearest tenth.

1. radius = __6 cm__

    diameter = __12 cm__

    circumference = __37.7 cm__

    height of cylinder = __18 cm__

    area of lateral surface = __678.6 cm²__

    area of 1 base = __113 cm²__

    area of 2 bases = __226 cm²__

    surface area = __904.6 cm²__

2. radius = __10 ft__

    diameter = __20 ft__

    circumference = __62.8 ft__

    height of cylinder = __30 ft__

    area of lateral surface = __1,884 ft²__

    area of 1 base = __314 ft²__

    area of 2 bases = __628 ft²__

    surface area = __2,512 ft²__

TAKE ANOTHER LOOK R105

Name _____

LESSON 25.3

Volume of Prisms and Pyramids

A pyramid can be thought of as part of a prism.
- You find the volume of a prism by multiplying length × width × height.
- You find the volume of a pyramid by multiplying $\frac{1}{3}$ × length × width × height.

A cube is a special prism whose length, width, and height have equal measures. In the cube below, length = width = height = 6 in.

$V = lwh = 6 \times 6 \times 6 = 216$
The volume is 216 in.^{3}.

The pyramid below has the same base and height as the cube. The pyramid has $\frac{1}{3}$ the volume of the cube.

$V = \frac{1}{3}bh = \frac{1}{3} \times 6 \times 6 \times 6 = 72$
The volume is 72 in.^{3}.

For Exercises 1–4, draw a prism and a pyramid with the given dimensions. Then find the volume of each.

1. $l = w = h = 9$ m

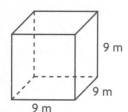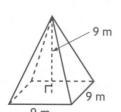

volume of cube = __729 m³__

volume of pyramid = __243 m³__

2. $l = 7$ cm; $w = 5$ cm; $h = 6$ cm

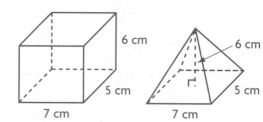

volume of prism = __210 cm³__

volume of pyramid = __70 cm³__

3. $l = w = h = 15$ ft

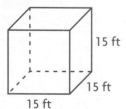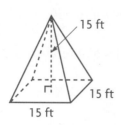

volume of cube = __3,375 ft³__

volume of pyramid = __1,125 ft³__

4. $l = w = h = 10$ m

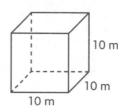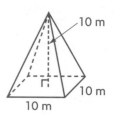

volume of cube = __1,000 m³__

volume of pyramid = __333 m³__

Name _____

LESSON 25.4

Volume of Cylinders and Cones

A cone can be thought of as part of a cylinder.
- You find the volume of a cylinder by multiplying B, the area of the base of the cylinder, times h, the height of the cylinder.
- You find the volume of a cone by multiplying $\frac{1}{3} \times B \times h$.

The cylinder below has a base with a radius of 7 cm and a height of 20 cm.

$V = \pi r^2 h = 3.14 \times 7^2 \times 20 = 3{,}077$
The volume is 3,077 cm³.

The cone below has a base with a radius of 7 cm and a height of 20 cm. The volume of the cone is $\frac{1}{3}$ the volume of the cylinder.

$V = \frac{1}{3}\pi r^2 h = \frac{1}{3} \times 3.14 \times 7^2 \times 20 = 1{,}026$
The volume is 1,026 cm³.

For Exercises 1–4, draw a cylinder and cone with the given radius and height. Then find the volume of each. Use $\pi = 3.14$. Round to the nearest tenth.

1. radius = 8 ft; height = 10 ft

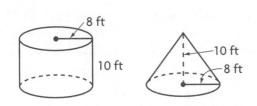

volume of cylinder = __2,009.6 ft³__

volume of cone = __669.9 ft³__

2. radius = 6 m; height = 24 m

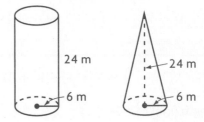

volume of cylinder = __2,713 m³__

volume of cone = __904.3 m³__

3. radius = 15 in.; height = 120 in.

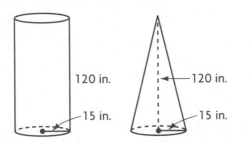

volume of cylinder = __84,780 in.³__

volume of cone = __28,260 in.³__

4. radius = 19 ft; height = 19 ft

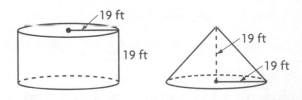

volume of cylinder = __21,537.3 ft³__

volume of cone = __7,179.1 ft³__

Name _____

LESSON 26.1

Changing Areas

Look for a pattern in the figures below.
Each rectangle has a perimeter of 16 ft.
Notice that as the length and width change, so does the area.

1 ft × 7 ft	2 ft × 6 ft	3 ft × 5 ft	4 ft × 4 ft	5 ft × 3 ft	6 ft × 2 ft	7 ft × 1 ft
area = 7 ft²	area = 12 ft²	area = 15 ft²	area = 16 ft²	area = 15 ft²	area = 12 ft²	area = 7 ft²

- What is the pattern? The rectangles that follow the square have the same dimensions as the rectangles that come before the square.
- Also, notice that the area is greatest when the rectangle is a square.

The greatest possible area of a rectangle with a given perimeter is often not a whole number.

Each rectangle below has a perimeter of 10 in.
The greatest area is 6.25 in.²

Divide to find the length of each side of a square.
perimeter ÷ 4 = 10 in. ÷ 4 = 2.5 in.

0.5 in. × 4.5 in.	1 in. × 4 in.	1.5 in. × 3.5 in.	2 in. × 3 in.	2.5 in. × 2.5 in.	3 in. × 2 in.
area = 2.25 in.²	area = 4 in.²	area = 5.25 in.²	area = 6 in.²	area = 6.25 in.²	area = 6 in.²

---

1. Complete the table for rectangles with perimeters of 14 ft.

Length (in ft)	2.0	2.5	3.0	3.5	4.0	4.5
Width (in ft)	5.0	4.5	4.0	3.5	3.0	2.5
Area (in ft²)	10.0	11.25	12	12.25	12	11.25

2. Draw each rectangle described in the table in Exercise 1.   1 unit = 1 ft
   Check students' drawings.

In Exercises 3–5, the perimeter of a rectangle is given. Find the length and width that will give the rectangle with the largest possible area. (HINT: Remember to divide.)

3. 25 m   4. 80 yd   5. 15 m

    __6.25 m × 6.25 m__   __20 yd × 20 yd__   __3.75 m × 3.75 m__

TAKE ANOTHER LOOK

Name _____

LESSON 26.2

Making Changes with Scaling

You can reduce or enlarge the sides of any figure by scaling. Write the scale percentage as a decimal, and then multiply.

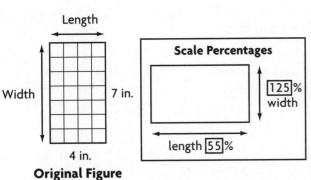
Original Figure

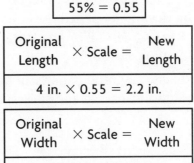

$$125\% = 1.25$$
$$55\% = 0.55$$

Original Length × Scale = New Length

4 in. × 0.55 = 2.2 in.

Original Width × Scale = New Width

7 in. × 1.25 = 8.75 in.

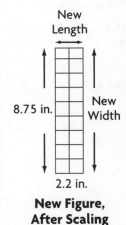
New Figure, After Scaling

For Exercises 1–6, use the given scale percentages to find the new length and width of each rectangle.

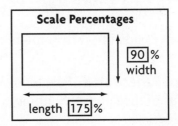

1. original length: 60 cm
   original width: 45 cm

   new length: __105 cm__

   new width: __40.5 cm__

2. original length: 10 in.
   original width: 10 in.

   new length: __17.5 in.__

   new width: __9 in.__

3. original length: 35 in.
   original width: 18 in.

   new length: __61.25 in.__

   new width: __16.2 in.__

4. original length: 24 in.
   original width: 54 in.

   new length: __42 in.__

   new width: __48.6 in.__

5. original length: 75 cm
   original width: 50 cm

   new length: __131.25 cm__

   new width: __45 cm__

6. original length: 9 in.
   original width: 7 in.

   new length: __15.75 in.__

   new width: __6.3 in.__

7. Use the dimensions of the original rectangle and the scale percentages to find the new length and width. Then draw the new figure.

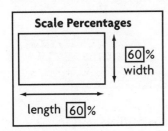

original length: 20 in.
original width: 10 in.

new length: __12 in.__

new width: __6 in.__

Draw the new figure.
1 unit = 1 in.
Check students' drawings.

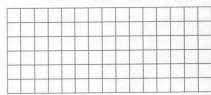

TAKE ANOTHER LOOK   R109

Name _____

LESSON 26.3

Problem-Solving Strategy

Making Models: Volume and Surface Area

Materials needed: graph paper, straightedge, scissors, tape

Pete's Pencil Company packages pencils in rectangular boxes that are 4 in. × 4 in. × 8 in. How many pencil boxes fit inside a shipping carton that is 8 in. × 8 in. × 16 in.? Find the surface area of one pencil box and one shipping carton.

- Copy each net below on graph paper. Cut out nets to make 1 model of the shipping carton and several models of the pencil box.

Shipping Carton 16 in. 8 in. 8 in.
1 Unit = 2 in.

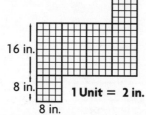

8 in. 4 in. **Pencil Box**

Put the pencil box models inside the carton model. The shipping carton holds 8 pencil boxes.

- Find the cardboard surface area of each box and the carton by adding the areas of the sides. Remember to include the top and bottom sides.

pencil box: $32 + 32 + 32 + 32 + 16 + 16 = 4 \times 32 + 2 \times 16 = 160$

shipping carton: $128 + 128 + 128 + 128 + 64 + 64 = 4 \times 128 + 2 \times 64 = 640$

So, the surface area of a pencil box is 160 in.^{2}. The surface area of a shipping carton is 640 in.^{2}.

For Exercises 1–4, make a model.

1. How many juice-glass boxes fit inside one shipping carton?

    _____ **32 juice-glass boxes** _____

2. What is the surface area of a juice-glass box?

    _____ **32 in.^{2}** _____

3. What is the surface area of the shipping carton?

    _____ **320 in.^{2}** _____

4. It costs $1\frac{1}{4}$ cents per square inch to manufacture each box. What does it cost to manufacture the juice-glass box? the shipping carton?

    _____ **$0.40; $4** _____

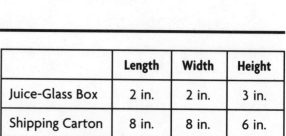

	Length	Width	Height
Juice-Glass Box	2 in.	2 in.	3 in.
Shipping Carton	8 in.	8 in.	6 in.

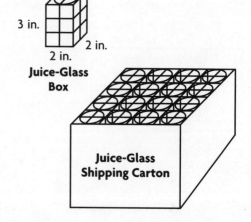

R110 **TAKE ANOTHER LOOK**

Name _____

LESSON 26.4

Volumes of Changing Cylinders

Original Juice now markets a new size—Jr. Original Juice.
The can for Jr. Original Juice is a model of the
Original Juice can made with a scale factor of $\frac{4}{5}$.
What are the dimensions of the Original Juice
can? How much juice does each can contain?

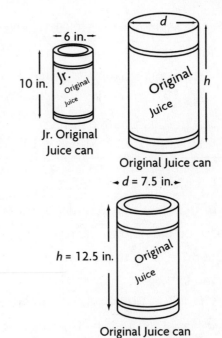

- Use a proportion and the scale factor to find
  the dimensions of the Original Juice can.

$\frac{4}{5} = \frac{\text{height of Jr. can}}{\text{height of Original can}}$ | $\frac{4}{5} = \frac{\text{diameter of Jr. can}}{\text{diameter of Original can}}$

$\frac{4}{5} \bowtie \frac{10}{h}$ | $\frac{4}{5} \bowtie \frac{6}{d}$

$4 \times h = 10 \times 5$ | $4 \times d = 6 \times 5$

$4h = 50$ | $4d = 30$

$\frac{4h}{4} = \frac{50}{4}$ | $\frac{4d}{4} = \frac{30}{4}$

$h = 12.5$ in. | $d = 7.5$ in.

- Use the volume formula $V = \pi r^2 h$ to find how much juice each can contains.

| V = volume | r = radius |
| $\pi \approx 3.14$ | h = height |

Jr. Original Juice can $r = 3$ in.   Original Juice can $r = 3.75$ in.

$V \approx 3.14\,(3)^2 \times 10$ $V \approx 3.14\,(3.75)^2 \times 12.5$

$ \approx 3.14\,(9) \times 10$ $ \approx 3.14\,(14.0625) \times 12.5$

$ \approx 282.6$ $ \approx 551.95312$

$V \approx 283$ in.^{3} $V \approx 552$ in.^{3}

Cylinder A   Cylinder B   Original Cylinder

For Exercises 1–3, use the cylinders shown.

1. Cylinder A was made using a scale factor of $\frac{1}{3}$. Find the dimensions of the original cylinder.

   $\frac{1}{3} = \frac{5}{h}$ $\frac{1}{3} = \frac{4}{d}$

   height = __15 in.__   diameter = __12 in.__

2. Cylinder B was made using a scale factor of $\frac{2}{3}$. Find the dimensions of cylinder B from the dimensions of the original cylinder.

   $\frac{2}{3} = \frac{h}{\boxed{15}}$ $\frac{2}{3} = \frac{d}{\boxed{12}}$

   height = __10 in.__   diameter = __8 in.__

3. What is the volume of cylinder A? of cylinder B? of the original cylinder?

   __62.8 in.^{3}; 502.4 in.^{3}; 1,695.6 in.^{3}__

TAKE ANOTHER LOOK R111

Name _____

LESSON 27.1

Percent and Sales Tax

The price of a computer is $1,499. What is the total cost if the sales tax rate is 7%?

- You know:
  total cost = (100% × price) + (7% × price)
  total cost = (100% × $1,499) + (7% × $1,499)
- Use the Distributive Property.
  (100% + 7%) × $1,499 = 107% × $1,499
- Write the percent as a decimal and multiply.
  1.07 × $1,499 = $1,603.93

You *multiplied* by 1.07 to find that the total cost equals $1,603.93.

Suppose you had known the total cost, $1,603.93, and the sales tax rate of 7%.
Find the price, p, of the computer.

- You know:
  (100% × price) + (7% × price) = total cost = $1,603.93
- Use the Distributive Property.
  (100% + 7%) × price = 107% × p = $1,603.93
- Write the percent as a decimal. $1.07p = \$1,603.93$
- Divide to solve for p. $\dfrac{1.07}{1.07}p = \dfrac{\$1,603.93}{1.07} = \$1,499$
- You *divided* by 1.07 to find that the price equals $1,499.

Find the total cost. Round to the nearest cent when necessary.

1. price: $40.00; tax rate: 6% __$42.40__
2. price: $56.70; tax rate: 5% __$59.54__
3. price: $350.00; tax rate: 8% __$378__
4. price: $76.55; tax rate: 7% __$81.91__
5. price: $49.99; tax rate: 3.5% __$51.74__
6. price: $325; tax rate: 6.5% __$346.13__

Find the price. Round to the nearest cent when necessary.

7. total cost: $420.00; tax rate: 5% __$400__
8. total cost: $265; tax rate: 6% __$250__
9. total cost: $105.93; tax rate: 7% __$99__
10. total cost: $64.79; tax rate: 8% __$59.99__
11. total cost: $5.69; tax rate: 3.5% __$5.50__
12. total cost: $136.25; tax rate: 9% __$125__
13. total cost: $338.52; tax rate: 4% __$325.50__
14. total cost: $0.83; tax rate: 6% __$0.78__
15. total cost: $52.70; tax rate: 5.5% __$49.95__
16. total cost: $93.09; tax rate: 7% __$87__

TAKE ANOTHER LOOK

Name _____

LESSON 27.2

Percent and Discount

RFG Warehouse has a clearance sale with a discount of 25% on all clothing.

You know the sale price of a jacket is $65.25. You can find the regular price, p.

- Subtract. $\quad\quad\quad\quad 100\% - 25\% = 75\%$
  The sale price is 75% of the regular price.
- Write an equation. $\quad\quad 75\% \times p = 65.25$
- Write the percent as a decimal. $0.75p = 65.25$
- Divide to solve. $\quad\quad \dfrac{0.75p}{0.75} = \dfrac{\$65.25}{0.75} = \$87.00$

So, the regular price of the jacket is $87.00.

Find the regular price. Round to the nearest cent.

1. sale price: $40.00; discount: 20%

    $50

2. sale price: $26.91; discount: 10%

    $29.90

3. sale price: $179.00; discount: 15%

    $210.59

4. sale price: $17.20; discount: 20%

    $21.50

5. sale price: $17.97; discount: 40%

    $29.95

6. sale price: $113.83; discount: 20%

    $142.29

7. sale price: $117.00; discount: 35%

    $180

8. sale price: $14.17; discount: 5%

    $14.92

9. sale price: $177.30; discount: 40%

    $295.50

10. sale price: $17.50; discount: 50%

    $35

11. sale price: $20.61; discount: 10%

    $22.90

12. sale price: $16.28; discount: 12%

    $18.50

13. sale price: $39.96; discount: 20%

    $49.95

14. sale price: $8.47; discount: 15%

    $9.96

15. sale price: $24.85; discount: 30%

    $35.50

16. sale price: $127.66; discount: 35%

    $196.40

17. sale price: $299.00; discount: 30%

    $427.14

18. sale price: $49.95; discount: 40%

    $83.25

19. sale price: $22.50; discount: 20%

    $28.13

20. sale price: $90.00; discount: 40%

    $150

TAKE ANOTHER LOOK

Name _____

LESSON 27.3

Percent and Markup

Percent means parts out of 100. To write a decimal as a percent, you multiply by 100, or move the decimal point two places to the right, and write %.

$0.16 = 16\%$ $0.148 = 14.8\% \approx 15\%$ $0.893 = 89.3\% \approx 89\%$

A store purchases computers at a *wholesale price* of $400.00 each and sells them at a *retail price* of $1,000.00 each. Find the value of the markup and the percent of markup.

retail price − wholesale price = markup

$\$1,000 - \$400 = \$600$

$\dfrac{\text{markup}}{\text{wholesale price}} = $ percent of markup

$\dfrac{600}{400} = 1.50 = 150\%$

The markup is $600 and the percent of markup is 150%.

In Exercises 1–16, the first amount is the wholesale price and the second amount is the retail price. Find the value of the markup and the percent of markup. Round to the nearest percent.

1. jeans: $20.00, $35.00

   $15; 75%

2. shirt: $16.00, $23.52

   $7.52; 47%

3. computer: $300.00, $400.00

   $100; 33%

4. shoes: $40.00, $80.00

   $40; 100%

5. pen: $0.50, $1.10

   $0.60; 120%

6. calculator: $8.00, $10.00

   $2; 25%

7. washing machine: $280.00, $420.00

   $140; 50%

8. pencils: $0.75, $0.98

   $0.23; 31%

9. necklace: $90.00, $270.00

   $180; 200%

10. shirt: $15.00, $30.00

    $15; 100%

11. toothbrush: $0.55, $1.30

    $0.75; 136%

12. eraser: $0.04, $0.05

    $0.01; 25%

13. car: $9,760.00, $11,614.00

    $1,854; 19%

14. postcard: $0.45, $0.60

    $0.15; 33%

15. book: $12.10, $21.60

    $9.50; 79%

16. notebook: $0.85, $1.25

    $0.40; 47%

17. For which Exercises is the percent of markup 100% or greater?

   Exercises 4, 5, 9, 10, and 11

R114 TAKE ANOTHER LOOK

Name _____

LESSON 27.4

Earning Simple Interest

The interest rate that banks pay is usually less than 10%.
Sometimes the interest rate has a mixed number, such as $5\frac{1}{2}\%$ or $7\frac{1}{4}\%$.

- To write the percent as a decimal,
  first write the mixed number as $5\frac{1}{2}\% = 5.5\%$ $7\frac{1}{4}\% = 7.25\%$
  a decimal.

- Move the decimal point 2 places to   05.5%   07.25%
  the *left*, and *remove the % symbol*.   0.055   0.0725

  $5\frac{1}{2}\% = 0.055$ $7\frac{1}{4}\% = 0.0725$

Find the interest earned on $650 of principal at a simple interest rate of $4\frac{3}{4}\%$
for 2 years.

- First, write the mixed number
  percent as a decimal. $4\frac{3}{4}\% = 4.75\% = 0.0475$

- Then use the formula: interest = principal × rate × time
                                              = 650 × 0.0475 × 2 = 61.75

The interest is $61.75.

For Exercises 1–14, write the mixed number percent as a decimal
percent and then as a decimal. Find the amount of interest earned.

1. $100 at $3\frac{1}{2}\%$ simple interest for 1 year

   _____3.5%, 0.035; $3.50_____

2. $2,000 at $6\frac{1}{4}\%$ simple interest for 2 years

   _____6.25%, 0.0625; $250_____

3. $200 at $3\frac{1}{4}\%$ simple interest for 2 years

   _____3.25%, 0.0325; $13_____

4. $225 at $6\frac{3}{4}\%$ simple interest for 4 years

   _____6.75%, 0.0675; $60.75_____

5. $350 at $5\frac{1}{2}\%$ simple interest for 3 years

   _____5.5%, 0.055; $57.75_____

6. $550 at $7\frac{1}{4}\%$ simple interest for 4 years

   _____7.25%, 0.0725, $159.50_____

7. $500 at $4\frac{1}{4}\%$ simple interest for 1 year

   _____4.25%, 0.0425; $21.25_____

8. $2,500 at $4\frac{1}{4}\%$ simple interest for 5 years

   _____4.25%, 0.0425, $531.25_____

9. $600 at $6\frac{3}{4}\%$ simple interest for 5 years

   _____6.75%, 0.0675; $202.50_____

10. $5,000 at $5\frac{3}{10}\%$ simple interest for 3 years

    _____5.3%, 0.053; $795_____

11. $1,000 at $5\frac{3}{4}\%$ simple interest for 1 year

    _____5.75%, 0.0575; $57.50_____

12. $610 at $5\frac{1}{2}\%$ simple interest for 4 years

    _____5.5%, 0.055; $134.20_____

13. $150 at $5\frac{1}{4}\%$ simple interest for 2 years

    _____5.25%, 0.0525; $15.75_____

14. $99 at $5\frac{9}{10}\%$ simple interest for 10 years

    _____5.9%, 0.059; $58.41_____

TAKE ANOTHER LOOK

Name _____

LESSON 27.5

Problem-Solving Strategy

Making a Table to Find Interest

Darell made a purchase of $230.57 on his credit card. The interest rate is 15%, or 1.25% monthly. He plans to make a payment of $50.00 each month. To help him compute and analyze his interest and balance amounts, Darell organizes the data in a table called a *spreadsheet*.

Interest is charged on the new balance. So, the amount in cell B3 is cell E2 × 0.0125. $180.57 × 0.0125 = $2.26

Always round interest up to the next cent.
Add the interest to the last balance. So, the amount in cell C3 is cell E2 + cell B3.
  $180.57 + $2.26 = $182.83

Complete the spreadsheet. Remember to round interest up to the next cent.

	A	B	C	D	E
1	Month	Interest (1.25%)	Balance	Payment	New Balance
2	July	$0	$230.57	$50.00	$180.57
3	August	$2.26	$182.83	$50.00	$132.83
4	September	$1.67	$134.50	$50.00	$84.50
5	October	$1.06	$85.56	$50.00	$35.56
6	November	$0.45	$36.01	$36.01	0
7	Total	$5.44		$236.01	

(1. row 4, 2. row 5, 3. row 6, 4. row 7)

Jennifer Hu charged $140.56 on her credit card. The interest rate is 18% annually, or 1.5% monthly. She plans to make a payment of $40.00 each month. Complete the spreadsheet.

	A	B	C	D	E
1	Month	Interest (1.5%)	Balance	Payment	New Balance
2	July	$0	$140.56	$40.00	$100.56
3	August	$1.51	$102.07	$40.00	$62.07
4	September	$0.94	$63.01	$40.00	$23.01
5	October	$0.35	$23.36	$23.36	0
6	Total	$2.80		$143.36	

(5. row 2, 6. row 3, 7. row 4, 8. row 5, 9. row 6)

R116 **TAKE ANOTHER LOOK**

Name _____

LESSON 28.1

Graphs and Pictures

Here are three graphs showing three possible relationships between speed and time.

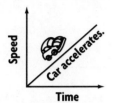 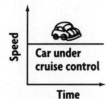 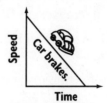

Speed increases as time increases.

Speed is constant (does not change) as time increases.

Speed decreases as time increases.

Think about the relationship between time and the height of a growing redwood tree. You can choose the graph below that best describes the relationship. Since a redwood tree grows with time, graph *b* best describes the relationship.

For Exercises 1–5, use the three graphs to the right showing relationships between distance and time. Choose the one that best describes the relationship given. Explain your choice.

1. the relationship between time and the distance a stalled car travels

   _____ c; distance does not change as the car sits still. _____

2. the relationship between time and your distance from school as you walk from home to school

   a; as time increases, your distance from

   school decreases.

3. the relationship between time and a train's distance from its point of origin as it travels

   b; as time increases, the distance

   from the point of origin increases.

4. the relationship between time and distance as a plane waits on the runway before takeoff

   _____ c; distance does not change as the plane waits. _____

5. the relationship between time and the distance between a basketball foul shot and the hoop

   a; as time increases, the distance from

   the hoop decreases.

a.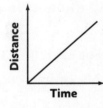

b.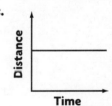

c.

TAKE ANOTHER LOOK R117

Name _____

LESSON 28.2

Relationships in Graphs

A graph tells a story.

A graph shows increases, decreases, or no change.

In the graph to the right, during what intervals did the number of cars sold increase?

- Look for the intervals where a segment rises.
- From 1986 to 1988 and from 1992 to 1994, the number of cars sold increased.

During what interval did the number of cars sold decrease?

- Look for the interval where a segment slopes down.
- From 1994 to 1996, the number of cars sold decreased.

During what interval did the number of cars sold remain constant?

- Look for the interval where a segment is horizontal.
- From 1988 to 1992, the number of cars sold remained constant.

The graph above tells *what* happened, but not *why* it happened.

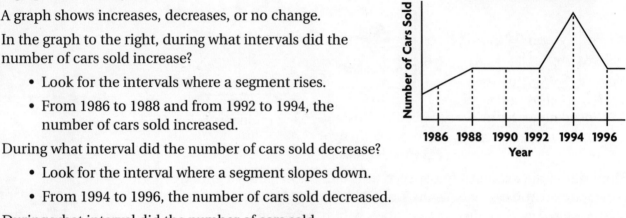

---

1. What reasons could you give for the increase in number of cars sold from 1986 to 1988 and from 1992 to 1994? __**Check students' answers.**__

2. What reasons could you give for the decrease in numbers of cars sold from 1994 to 1996? __**Check students' answers.**__

For Exercises 3–6, use the "Water in Reservoir" graph.

3. During what intervals did the water level increase?

   __**from 10 to 12 days**__

4. During what intervals did the water level remain constant?

   __**from 0 to 3 days and from 6 to 10 days**__

5. During what intervals did the water level decrease?

   __**from 3 to 6 days and from 12 to 14 days**__

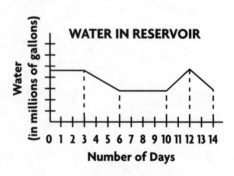

6. Use the information in the graph to tell a story about the water level in the reservoir. **Check students' stories.**

_____

_____

R118 **TAKE ANOTHER LOOK**

Name _____

LESSON 28.3

Graphing Relationships

A graph illustrates a relationship.

Make a graph to show the relationship: The more you earn, the more you save.

Step 1 • What are the variables? *earn* and *save*
• So, label the axes *Earnings* and *Savings*.

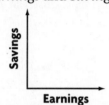

Step 2 • The words "more" and "more" tell you that the relationship is increasing.

Step 3 • Sketch a graph.

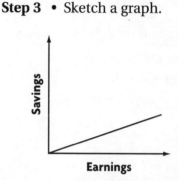

In Exercises 1–4, identify the variables, label the axes, and sketch a graph of the relationship.

1. **Relationship:** The price of bread has not changed during the last 2 months.

   Step 1 The variables are __price__ and __time__.

   Step 2 "Not changed" means the graph will do which of the following: increase, decrease, or remain constant?

   __remain constant__

   Step 3 Sketch the graph.

2. **Relationship:** Your science class drops a ball from the roof of the school building. The ball's distance from the ground decreases as time increases.

   Step 1 The variables are __distance__ and __time__.

   Step 2 "Decreases" means the graph will do which of the following: increase, decrease, or remain constant?

   __decrease__

   Step 3 Sketch the graph.

3. **Relationship:** The greater the number of options on a car, the higher the cost.

   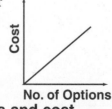

   __number of options and cost__

4. **Relationship:** The later I start my homework, the less homework I can do.

   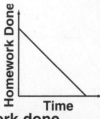

   __time and homework done__

TAKE ANOTHER LOOK R119

Name _____

LESSON 28.4

Using Scatterplots

Is there a relationship between the speed of a car and the distance it takes to stop once the brakes are applied?

Speed (in mph)	10	20	30	40	50	60	70
Stopping Distance (in ft)	20	40	70	100	150	230	300

A *scatterplot* can show you.

Step 1 Use the table to name the ordered pairs.

(speed, distance)
(10,20)
(20,40)
(30,70)
(40,100)
(50,150)
(60,230)
(70,300)

Step 2 Label the axes. Graph the ordered pairs as a scatterplot.

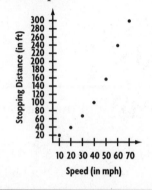

Step 3 Describe the relationship as *a positive, a negative,* or *no* correlation.

- Since the points appear to describe a line, there is a correlation.
- Since the line slants up, there is a positive correlation.

Use the table to name ordered pairs. Graph the ordered pairs. Then tell if the relationship shows *a positive, a negative,* or *no* correlation.

Year	5	6	7	8	9	10
Whistles Produced (in millions)	77.9	78.5	79.1	83.0	82.4	83.8

Step 1 Name the ordered pairs.

(year, whistles produced)

(__5__ , __77.9__)

(__6__ , __78.5__)

(__7__ , __79.1__)

(__8__ , __83.0__)

(__9__ , __82.4__)

(__10__ , __83.8__)

Step 2 Graph the ordered pairs on a scatterplot. **Check students' graphs.**

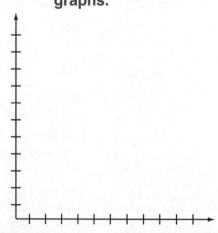

Step 3 The scatterplot shows __a positive__ correlation.

R120 **TAKE ANOTHER LOOK**